Contents

Preface

This book is intended as an introduction to differential equations. Although it has some mathematical definitions and theorems it focuses as a more practical guide for early university students. Each section will start by looking at some basic theory before going on to looking at a range of examples. The reader should pay careful attention to these as they were carefully chosen to give a wider spectrum of the material.

We will look at how to solve first and second order ordinary differential equations, and will touch on higher order equations towards the end of the book. Each section uses a range of methods and plots solutions graphically. We interpret these graphical solutions to understand further what our algebraic solutions are.

Exercises and solutions (in the rear of the book) accompany each chapter to test the skills learnt and cement understanding. Of particular interest to engineering students this book looks at the method of Laplace transforms to solve ODEs in chapter 4. These are also of interest to mathematicians and physicists and appear in a variety of modules in these degrees.

Chapter 1 is a brief historical account and recap on basic calculus. The mathematical history would not be fundamental in a standard course on ODEs and the mathematical recap would have been covered within an a-level course. If the reader is already comfortable with these they can skip to chapter 2 without any key content being missed.

This book aims to take the theory learnt and apply this to real world examples to get a further and deeper understanding of the material. This does mean it is for a more applied mathematician as an audience. For a more pure background the reader may need to dive further into a wider selection of reading material.

Chapter 1

Introduction

Mankind has been concerned with the modeling of the real world since intelligent life began. This modeling could be as a basis of civilisation such as currencies or looking for a competitive edge such as faster cars, financial portfolios, or supply chain management. Primarily mathematics was built from a need to compute and branched off in many directions as the needs of civilization saw fit. Numbers systems were introduced to allow comparing quantities, early algebra came in to aid in the solving of problems using symbols in place of missing values, and geometry focused on shape, space and position. The Greeks started a trend of looking at the purity of mathematics with the rich and influential of the era dedicating their time to it as a hobby. This caused a split into pure and applied mathematics as it was no longer a need that drove it but an aesthetic love for the subject matter. This approach also linked mathematics with philosophy with the great thinkers of the time delving into both.

In mankind's quest to model the universe, it needed to consider that in the real world quantities change. One good example of this is throwing a ball in the air. The position of the ball x changes as time t passes. Calculus enables us to see how the ball moves and outline where it will be in the future with the given parameters. The example we used above is from mechanics but the rate of change is important in many other areas. A few examples are biology with the spread of illness as well as economics and changes in currency. Calculus, the study of continuous change is the foundation of what this book will focus on.

1.1 History

Calculus, translated from Latin, directly means small pebble. Pebbles were used as a counting device and, as we will discover later, the smaller the pebble the more accurate the calculations computed.

Calculus is split into two main areas; differential calculus, and integral calculus. These focus on rates of change and areas between curves respectively. The two areas are linked together with the fundamental theorem of calculus which states that differentiation and integration are inverses of each other.

Theorem 1: (Fundamental theorem of calculus) Let f be a continuous real-valued function on the closed interval [a,b]
i) Let

$$F(x) = \int_a^x f(s)ds$$

$\forall x \in [a, b]$ then

$$F'(x) = f(x).$$

ii) Let F(x) be the indefinite integral of f on [a,b] then

$$\int_a^b f(x)dx = F(b) - F(a).$$

What we understand as calculus now was built on the geometrical work by the Greeks and the algebraic work of the Descartes and Wallis. With this previous work, two further men battled over the discovery of calculus in its more recognisable form.

Issac Newton, born 25th December 1642 England, was the son of a freehold farmer. Unfortunately for Newton his farther died three months before his birth. Newton studied at Trinity college, Cambridge, paying his way initially before being awarded a scholarship in 1664. In 1669 Newton was appointed the position of Lucasian professor (a mathematics professorship at the University of Cambridge) one year after finishing his MA, where he stayed until 1696. After his time at Cambridge, Newton went on to be the master of the mint. Newtons work touched many areas but he is often remembered for his works in mechanics. He studied works from the great thinkers of previous times before putting these together for his future work. One of Newtons most notable discoveries and often miss quoted in popular media is the three laws of motion;

1. If the net force of an object is zero then the object has constant velocity.

2. The rate of change of the momentum of a body is directly proportion to the force applied.

3. For every action there is an equal and opposite reaction.

This mechanics background can be seen in his notation and the approach he used to develop his method of calculus. He was later knighted in 1705 by Queen Anne. Newton died in his sleep in 1727 and was buried at Westminster Abbey.

Gottfried Wilhelm Leibniz, born 1st July 1646 Germany towards the end of the Thirty Years War, was the son of Friedrich Leibniz and Catharina Schmuck. Friedrich had been a professor of philosophy at the University of Leipzig and clearly his aptitude for studies was passed down to his son.

Leibniz's early studies focused around philosophy and law. At the age of 14 he started at the University of Leibzig going on to achieve his Bachelors degrees and Masters degree there. He submitted a doctoral thesis which was turned down and he subsequently left the university. Leibniz earned his doctorate in Law in 1666 and was then able to practice.

Leibniz didn't start focusing on mathematics until 1672 when he traveled to Paris and met Christian Huygens. Enjoying the mathematics and wanting to broaden his understanding he studied under Huygens. Leibniz focused on many areas of mathematics such as linear systems, geometry and the early stages of topology.

Leibniz died in 1716. He was out of favour at the time due to his controversy with Newton over the discovery of calculus. Even though he was a life member of the Royal Society the institution was not represented at his funeral.

During Leibniz's study under Huygens and surrounded by Descartes and Pascal, he found his new calculus between 1673 and 1676 publishing between 1684 to 1686. Newton on the other hand did not publish until 1704 but claimed he had discovered between 1665 and 1666. Newton's only proof that he had discovered his calculus a year earlier was his word which at the time carried great influence and many did not want to doubt him. Many of those who believed Newton claimed that Leibniz had seen Newtons papers and developed his calculus from that. Some mathematicians such as L'Hopital gave credit to both men when referencing calculus but overall Leibniz fell out of favour with many. He died with his accomplishments still in question.

The current belief from historical mathematicians is that both men discovered it independently with Leibniz publishing first. Leibniz's notation is often preferred to Newtons. The methods of both lacked rigour and further constructive work built upon their ideas to produce the calculus we see today.

1.2 Differentiation

Now that we have a brief snapshot of the men discovered calculus let us take more detailed look at the mathematics itself. Of the two strains of calculus, differential and integral, our focus for the rest of the chapters lies on the former.

1.2.1 The Derivative

Let us start by taking a linear equation of a straight line for a graph;

$$y = 2x + 5$$

We know from secondary school mathematics that if we want to find out how steep the line is, the gradient, we need to take two points on the line and look at the difference. More specifically we take the difference in y and divide by the corresponding difference in x.

$$m = \frac{\Delta y}{\Delta x} \tag{1.1}$$

where m is the gradient and Δ is the difference.

From our equation 1.1 we can work out the gradient if we had two points (or recognise that the gradient is the coefficient of x when we have solved for y as is 2 for the equation above). However this is a concept only for linear equations. What happens when we want a similar concept for polynomials of higher powers?

For this let us look at the equation of the tangent of a curve at a certain point. This is the straight line that cuts a curve at a given point.

Let $y = x^2$. If we want to find the equation of the tangent at a given point x_0 we first need to find the gradient of the tangent by

$$m = \lim_{x \to x_0} \frac{f(x) - f(x_0)}{x - x_0} \tag{1.2}$$

and substitute this at the given point into the form for a linear equation (since the tangent is linear):

$$y - f(x_0) = m(x - x_0). \tag{1.3}$$

Using this formula and $x_0{=}1$ we obtain the gradient to be

$$
\begin{aligned}
m &= \lim_{x \to 1} \frac{x^2 - f(1)}{x - 1} \\
&= \lim_{x \to 1} \frac{x^2 - 1}{x - 1} \\
&= \lim_{x \to 1} \frac{(x - 1)(x + 1)}{x - 1} \\
&= \lim_{x \to 1} (x + 1) \\
&= 2
\end{aligned}
$$

Using this gradient we find the equation of the tangent to equal;

$$
\begin{aligned}
y - f(x_0) &= m(x - x_0) \\
y - 1 &= 2(x - 1) \\
y &= 2x - 1
\end{aligned}
$$

1.15 is the tangent of the line $y = x^2$ at $x_0 = 1$. Now if we take $x_0 = 2, 3$ we get the gradients to be

$$m = \lim_{x \to 2} \frac{x^2 - 4}{x - 2} = 4,$$

$$m = \lim_{x \to 3} \frac{x^2 - 9}{x - 3} = 6.$$

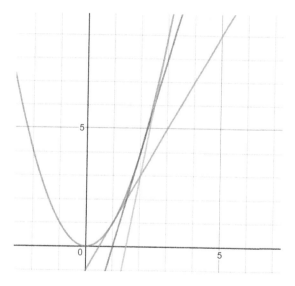

Figure 1.1: The graph of $y = x^2$ with tangents at x=1, x=2 and x=3

We can see that every time we increase our value of x_0 by 1, our tangents gradient increases by two, the same value as our power.

Let us now take our formula for the gradient of the tangent and make it more general. Let $h = x - x_0$.

$$m = \lim_{h \to 0} \frac{f(x_0 + h) - f(x_0)}{h}$$

You can check that this formula still gives the same values for the gradient of the tangents at the points we used above.

As we want to find an expression for how the gradient changes we can replace x_0 with x, our definition of the derivative, $f'(x)$,

$$f'(x) = \lim_{h \to 0} \frac{f(x + h) - f(x)}{h}.$$

Let us now calculate the derivative of $y = x^2$.

$$\begin{aligned} f'(x) &= \lim_{h \to 0} \frac{(x + h)^2 - x^2}{h} \\ &= \lim_{h \to 0} \frac{x^2 + 2hx + h^2 - x^2}{h} \\ &= \lim_{h \to 0} \frac{2hx + h^2}{h} \\ &= \lim_{h \to 0} 2x + h = 2x \end{aligned}$$

This agrees with what we saw earlier with the rate of change as x_0 increased by one. Let us now look at a more general function, that of a polynomial with degree n.

$$(x^n)' = \lim_{h \to 0} \frac{(x+h)^n - x^n}{h}$$
$$= \lim_{h \to 0} \frac{x^n + nhx^{n-1} + (n-1)h^2x^{n-2}... - x^n}{h}$$
$$= \lim_{h \to 0} (nx^{n-1} + (n-1)hx^{n-2}...)$$
$$= nx^{n-1}$$

This method can be used to find the derivative of any function but the common functions are usually committed to memory. A table below contains some of the important functions and their derivatives.

f(x)	f'(x)
A	0
x^n	nx^{n-1}
$sin(x)$	$cos(x)$
$cos(x)$	$-sin(x)$
e^{Ax}	Ae^{Ax}
$ln(x)$	$\frac{1}{x}$
A^x	$A^x ln(A)$

1.2.2 Product and Quotient Rule

When dealing with functions and derivatives of these we need to note that

$$\frac{d}{dx}[f(x)g(x)] \neq \frac{d}{dx}f(x)\frac{d}{dx}g(x). \tag{1.4}$$

To show this let us take $f(x) = x$ and $g(x) = 6x^2$. If we take derivatives of both of these we get $f'(x) = 1$ and $g(x) = 12x$ such that $f'(x)g'(x) = 12x$.

Now let us look at what we get for $(f(x)g(x))'$. $(f(x)g(x)) = 6x^3$ and thus taking the derivative gives $(f(x)g(x))' = 18x^2$. We clearly see that these two terms are not equal.

This is where the product and quotient rules becomes useful. They help us find derivatives of functions that multiply each other or are quotients of one another.

Formula: (Product rule) Let f(x) and g(x) be functions of x, then

$$\frac{d}{dx}(f(x)g(x)) = \frac{df}{dx}(x)g(x) + f(x)\frac{dg}{dx}(x). \tag{1.5}$$

Formula: (Quotient rule) Let f(x) and g(x) be functions of x, then

$$\frac{d}{dx}\left(\frac{f(x)}{g(x)}\right) = \frac{\frac{df}{dx}(x)g(x) - f(x)\frac{dg}{dx}(x)}{g(x)^2}. \tag{1.6}$$

We are only going to look at the proof of the product rule.

Proof:

Using the definition of the derivative we see that

$$\frac{d}{dx}(f(x)g(x)) = \lim_{h \to 0} \frac{f(x+h)g(x+h) - f(x)g(x)}{h}$$

We use the trick of adding and subtracting $f(x+h)g(x)$. This is a useful trick in mathematical proofs.

$$\frac{d}{dx}(f(x)g(x))$$
$$= \lim_{h \to 0} \frac{f(x+h)g(x+h) - f(x+h)g(x) + f(x+h)g(x) - f(x)g(x)}{h}$$

We can factorise these into two brackets, one with $f(x+h)$ outside the bracket and one with $g(x)$ outside the brackets.

$$\frac{d}{dx}(f(x)g(x))$$
$$= \lim_{h \to 0} \frac{f(x+h)[g(x+h) - g(x)] + g(x)[f(x+h) - f(x)]}{h}$$

Separating this into two fractions we see that it starts to look like the definitions of $f'(x)$ and $g'(x)$.

$$\frac{d}{dx}(f(x)g(x))$$
$$= \lim_{h \to 0} f(x+h) \frac{g(x+h) - g(x)}{h} + \lim_{h \to 0} g(x) \frac{f(x+h) - f(x)}{h}$$

We note the property of limit that when you take a limit of two functions that are multiplied together it is the same as taking the limit of each and multiplying them together. This gives

$$\frac{d}{dx}(f(x)g(x))$$
$$= \lim_{h \to 0} f(x+h) \lim_{h \to 0} \frac{g(x+h) - g(x)}{h} + \lim_{h \to 0} g(x) \lim_{h \to 0} \frac{f(x+h) - f(x)}{h}.$$

This gives, using the definition of derivatives,

$$\frac{d}{dx}(f(x)g(x)) = f'(x)g(x) + f(x)g'(x).$$

Example: Find $\frac{d}{dx}(sin(x)x^3)$.

Solution:

This example is of the form $\frac{d}{dx}[f(x)g(x)]$ so we can use the product rule. We let $f(x) = sin(x)$ and $g(x) = x^3$. Then we have

$\frac{df}{dx}(x) = cos(x)$ and $\frac{dg}{dx}(x) = 3x^2$. Substituting these into the product rule we obtain

$$\frac{d}{dx}(sin(x)x^3) = cos(x)x^3 + 3sin(x)x^2.$$

Example: Find $\frac{d}{dx}\frac{e^{2x}}{2x^2}$.

Solution:

In this example we have a fractions of functions of x that we are taking derivatives of. This means we can use the quotient rule. We let $f(x) = e^{2x}$ and $g(x) = 2x^2$. Then we have $\frac{df}{dx}(x) = 2e^{2x}$ and $\frac{dg}{dx}(x) = 4x$. Substituting these into the quotient rule gives

$$\frac{d}{dx}\frac{e^{2x}}{2x^2} = \frac{2x^2e^{2x} - 4xe^{2x}}{4x^4}.$$

1.2.3   Chain Rule

Often we do not just have a function on their own and we need to compute a derivative of a composite function.

Formula: (Chain rule) Let f(x) and g(x) be composite functions. Then

$$\frac{d}{dx}f(g(x)) = f'(g(x))g'(x). \tag{1.7}$$

The chain rule was first introduced by Leibniz with his work on calculus.

Example: Find $\frac{dy}{dx}$ of $y = (2x + 4)^3$.

Solution:

Let $u = 2x + 4$ such that $y = u^3$. The chain rule states that

$$\frac{dy}{dx} = \frac{dy}{du} \cdot \frac{du}{dx}.$$

Using $u = 2x + 4$ we can calculate the derivative of u with respect to x

$$\frac{du}{dx} = 2.$$

Calculating the derivative of y with respect to u

$$\frac{dy}{du} = 3u^2.$$

Substituting these back we obtain

$$\frac{dy}{dx} = 3u^2(2)$$

giving our final answer to be

$$\frac{dy}{dx} = 6(2x + 4)^2.$$

Example: Find $\frac{dy}{dx}$ of $y = e^{3x-1}$.

Solution:

Let $u = 3x - 1$ such that $y = e^u$. The chain rule states that

$$\frac{dy}{dx} = \frac{dy}{du} \cdot \frac{du}{dx}.$$

Using $u = 3x - 1$ we can calculate the derivative of u with respect to x

$$\frac{du}{dx} = 3.$$

Calculating the derivative of y with respect to u

$$\frac{dy}{du} = e^u.$$

Substituting these back we obtain

$$\frac{dy}{dx} = (3)e^u$$

giving our final answer to be

$$\frac{dy}{dx} = 3e^{3x-1}.$$

1.2.4 Implicit Differentiation

We may encounter equations which would be difficult to rearrange into a form in which one variable equals a function of another. This makes finding a derivative outright difficult too. Implicit differentiation can be used in this case. With implicit differentiation we note that

$$\frac{d}{dx}f(y) = \frac{d}{dy}f(y)\frac{dy}{dx}. \tag{1.8}$$

Example: Find $\frac{dy}{dx}$ of $y^2 + x^2 - 3 = 4x + y^3$.

Solution:

We start by differentiating each term with respect to x.

$$\frac{d}{dx}y^2 + \frac{d}{dx}x^2 - \frac{d}{dx}3 = \frac{d}{dx}4x + \frac{d}{dx}y^3.$$

The derivatives of the functions of x are trivial. For the derivatives of the functions of y we need to use equation 1.8. Together these give

$$2y\frac{dy}{dx} + 2x = 4 + 3y^2\frac{dy}{dx}.$$

Rearranging to get all the terms with the derivative onto the left hand side and factorising

$$\frac{dy}{dx}(2y - 3y^2) = 4 - 2x.$$

Diving both sides by $(2y - 3y^2)$ we obtain

$$\frac{dy}{dx} = \frac{4 - 2x}{2y - 3y^2}.$$

Example: Find $\frac{dy}{dx}$ of $x^4 + sin(y) = e^{2y}$.

Solution:

We start by differentiating each term with respect to x.

$$\frac{d}{dx}x^4 + \frac{d}{dx}sin(y) = \frac{d}{dx}e^{2y}$$

Again we are going to use equation 1.8

$$4x^3 + cos(y)\frac{dy}{dx} = 2e^{2y}\frac{dy}{dx}$$

Rearranging to get all the terms with the derivative onto the left hand side and factorising

$$\frac{dy}{dx}(2e^{2y} - cos(y)) = 4x^3.$$

Diving both sides by $2e^{2y} - cos(y)$ we obtain

$$\frac{dy}{dx} = \frac{4x^3}{2e^{2y} - cos(y)}.$$

1.3 Integration

Differentiation looked at the rate of change. We saw this with the derivative for an equation for a line. Like with addition and sub-traction being inverses, the inverse of differentiation is integration. This is stated in the fundamental there om of calculus. Graphically when integrating a curve over a domain we obtain the area between the curve and the axis. This will have finite domain over which we can obtain this area. If though we have this integral over the whole domain, in our case $\mathbb{R}$, then we call this an indefinite integral.

Like differentiation, we can memorise the key integrals and make use of the table in the previous chapter. However we will take time to look at a key skill called integration by parts and integration using substitution.

1.3.1 Integration by Parts

When we have an integral of a single function we can use tables to find the integral. We use the tables in reverse to do this. However if we have an integral to calculate that is made up of more than one function then these tables are not so useful. This is where integra-tion by parts comes in useful.

Formula: (Integration by parts) Let $u(x)$ and $v(x)$ be functions. Then

$$\int_a^b u(x)\frac{dv}{dx}(x)dx = \left[u(x)v(x)\right]_a^b - \int_a^b \frac{du}{dx}(x)v(x)dx \qquad (1.9)$$

What we see here is that the derivative passes from $v(x)$ to $u(x)$. This means that if $u(x)$ differentiates to a constant we will obtain an integral that we can do using tables (or by memory). In more advanced mathematics, the method of integration by parts is an im-portant analysis tool and can be used to great effect.

Let us look at an example.

Example: Find

$$\int 3x cos(x) dx.$$

Solution:

As mentioned above we can be clever about our choices of which function we call $u(x)$ and $v(x)$. Here we note that if we differentiate $3x$ we will obtain a constant meaning that in 1.9 the integral on the right hand side is one we want. Thus we will choice

$$u(x) = 3x \qquad\qquad \frac{dv}{dx}(x) = cos(x).$$

giving

$$\frac{du}{dx}(x) = 3 \qquad\qquad v(x) = sin(x).$$

Now we will substitute these into the integration by parts formula 1.9

$$\int 3x cos(x) dx = \big[3x sin(x)\big] - \int 3 sin(x) dx.$$

Finally we can perform the integration on the right hand side.

$$\int 3x cos(x) dx = 3x sin(x) + 3 cos(x) dx.$$

Example: Find

$$\int e^{4x} cos(x) dx.$$

Solution:

This example has one important difference from the first, neither of the two functions differentiate a finite number of times to give a constant. This may be problematic but we will see a trick that can help us solve this specific and similar examples.

We start as if it is an integration by parts problem and naming the integral Y. Here we have

$$u = e^{4x} \qquad\qquad dv = cos(x)$$
$$du = 4e^{4x} \qquad\qquad v = sin(x).$$

Thus we get

$$Y = e^{4x}sin(x) - 4\int e^{4x}sin(x)dx.$$

Again we have to use integration by parts with

$$u = e^{4x} \qquad\qquad dv = sin(x)$$
$$du = 4e^{4x} \qquad\qquad v = -cos(x)$$

giving

$$Y = e^{4x}sin(x) + 4e^{4x}cos(x) - 16\int e^{4x}cos(x)dx.$$

What we see here is that the integral on the right hand side is actually Y. By substituting this in place of the integral we can solve for Y.

$$Y = e^{4x}sin(x) + 4e^{4x}cos(x) - 16Y$$

For the final solution we get

$$Y = \frac{1}{17}e^{4x}\big(sin(x) + 4cos(x)\big).$$

The last two examples did not have limits of integration. This made the calculations shorter than if we had limits as we were not required to substitute our limits to obtain our final answer. Let us look at one more example where we do have limits of integration.

Example: Find

$$\int_1^3 2xe^{4x}\,dx.$$

Solution:

As we did before we can make a choice for our $u(x)$ and $v(x)$.

$$u(x) = 2x \qquad\qquad \frac{dv}{dx}(x) = e^{4x}.$$

giving

$$\frac{du}{dx}(x) = 2 \qquad\qquad v(x) = \frac{1}{4}e^{4x}.$$

Using 1.9 we obtain

$$\int_1^3 2xe^{4x}\,dx = \left[\frac{1}{2}xe^{4x}\right]_1^3 - \int_1^3 \frac{1}{2}e^{4x}\,dx.$$

Performing the integration on the right hand side

$$\int_1^3 2xe^{4x}\,dx = \left[\frac{1}{2}xe^{4x}\right]_1^3 - \left[\frac{1}{8}e^{4x}\right]_1^3.$$

This is where this example differs from the last ones. We are not finished yet as we still have to substitute our limits of integration. This means replacing our x in the bracket by the opt value and subtracting the bracket whilst replacing it with the bottom value. Doing this substitution yields

$$\int_1^3 2xe^{4x}\,dx = \frac{1}{8}\left[11e^{12} - 3e^4\right].$$

1.3.2   Integration by Substitution

Another more specific method of integration is substitution. We can use this method when taking the integral of the product of two functions where one is a derivative of the other.

$$\int f(g(x))g'(x)dx \qquad\qquad (1.10)$$

If we make the substitution of $u = g(x)$ then

$$\frac{du}{dx} = g'(x)$$
$$du = g'(x)dx$$

by chain rule. Using this we can make 1.10

$$\int f(u)du. \qquad\qquad (1.11)$$

Example: Find

$$\int 10xe^{5x^2}dx.$$

Solution:

Here we have $g(x) = 5x^2$ so we make the substitution of $u = 5x^2$ and $du = 10xdx$. Making this substitution we get

$$\int e^u du = e^u + c.$$

Once again using $u = 5x^2$ we obtain our final solution

$$\int 10xe^{5x^2}dx = e^{5x^2} + c$$

1.4 The Differential Equation

We are now going to take a brief history of differential equations.

Definition: A differential equation is an equation which contains one or more derivative in it.

Newton's work on mechanics gave birth to some of the earliest differential equations however the first use of the term 'Differential Equation' was by William Emerson in 1763. Newton solved a lot of the differential equations he found without the use of his calculus, and used graphical methods. In his work 'Methodus fluxionum et Serierum Infinitarum', 1671, Newton listed three differential equations;

1.

$$\frac{dy}{dx} = f(x)$$

2.

$$\frac{dy}{dx} = f(x, y)$$

3.

$$x_1 \frac{\delta y}{\delta x_1} + x_2 \frac{\delta y}{x_2} = y$$

where the third is known as a partial differential equation. He solved these using infinite series.

A lot of the early work of differential equations were done between Leibniz and the Bernoulli brothers. In letters to Leibniz, Bernoulli proposed a new differential equation in the form

$$\frac{dy}{dx} + P(x)y = Q(x)y^n. \tag{1.12}$$

This is named after Bernoulli and we will look at how to solve these style of equations. Leibniz solved these in 1696.

Numerical methods can also be used for solving differential equations. Euler was one of the first to publish with his method in 1768 with Cauchy proving convergence in 1824. Runge and Kutta put a lot of work into solving using methods named after them.

1.5   Real World Differential Equations

As we discussed earlier, differential equations are extremely important in modeling our world mathematically. Let us now take a look at a few examples.

1.5.1   Newton's Second Law

We are going to continue our use of mechanical systems for our examples with the use of Newton's second law $F = ma$ wtih force F, mass m and acceleration a. We can express the acceleration as either a first or second order derivative;

$$F = m\frac{dv}{dt}$$
$$F = m\frac{d^2s}{dt^2}$$

with velocity v and displacement x.

Newton's second law can be applied to many mechanical situations. One such systems is a mass on a spring being pulled down by gravity.

We have mass m, acceleration due to gravity g, spring constant k, extension due to the downward force s, and the extension at a given point x. We will take the extension downward as the positive direction.

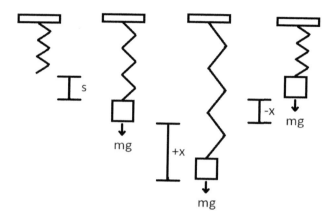

Figure 1.2: Spring extension diagrams.

The force acting downward, in the positive direction is mg. The force acting upward, in the negative direction is $-ks - kx$. This gives

$$mg - ks - kx = m\frac{d^2x}{dt^2}.$$

If we take the body at equilibrium then we get $mg - ks = 0$ so

$$m\frac{d^2x}{dt^2} = -kx.$$

These differential equations mostly give solutions that are harmonic (or more bluntly, they contain the sine and cosine functions). We call the movement produced by these functions harmonic motion.

1.5.2   Population Growth

Another example is population growth. The population of humans, plants, animals, and bacteria can be modelled over time by using a first order differential equation. If we assume there is no inhibiting factor such as the environment, which at low populations is not a major influence, then we see the growth is proportional to its size. If we let $y = y(t)$ be the size of the population at time t the rate of increase of the population will be $\frac{dy}{dt}$ giving us the equation

$$\frac{dy}{dt} = Ay \qquad (1.13)$$

where A is the growth rate coefficient. y(0), our initial condition (see chapter 2 for these), is the starting population. We can make a plot of the solution to this differential equation and use a starting population, $y(0) = 1000$.

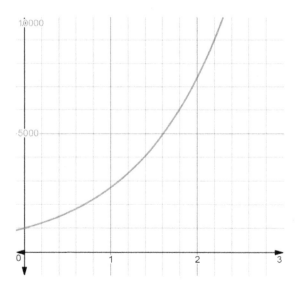

Figure 1.3: The graph of uninhibited population growth with a starting population of 1000.

This model may not hold at higher populations as the environment may not be able to cater for the larger population and at this point we call it the carrying capacity, L. Belgium mathematician Verhulst came up with a different model for population growth under this restriction which is called the logistical model. The differential equation associated with this model is

$$\frac{dy}{dx} = A(1 - \frac{y}{L})y. \tag{1.14}$$

1.5.3 Interest Rates

Interest rates on bank accounts provide us with another use of the differential equal. To calculate compound interest on a bank account we use the following formula

$$M = M_0(1 + \frac{r}{n})^{nt} \tag{1.15}$$

where M is the amount of money at a given time, M_0 is the initial amount of money, r is the interest rate, n is the number of times interest is paid per year, and t is the number of years.

If we let the number of time that interest is paid tend to infinity we get what is known as continuously compound interest.

$$\lim_{n \to \infty} M_0(1 + \frac{r}{n})^{nt}$$

Let $x = \frac{n}{r}$ so $n = rx$. Substituting this we get

$$M_0 \lim_{x \to \infty} (1 + x)^{xrt}$$

$$= M_0 \lim_{x \to \infty} ((1 + x)^x)^{rt}.$$

The $\lim_{x \to \infty}((1 + x)^x) = e$ so we obtain our general solution

$$y = M_0 e^{rt}.$$

1.6   Exercises

1. Find the gradient of the line that passes through the points (3,4) and (6,5).

2. Find the gradient of the line that passes through the points (6,1) and (2,9).

3. Is there a single straight line that passes through (2,1), (-1,0), and (5,4)?

4. Find the equation of the tangent to $y = 2/x$ at $x_0 = 2$.

5. Prove that the derivative of e^x is e^x

6. Use your answer to question 3 and a substitution of $u = nx$ to prove that the derivative of e^{nx} is ne^{nx}

7. Find $\frac{dy}{dx}$ of
$$y = (2x^3 + 5)^4.$$

8. Find $\frac{dy}{dx}$ of
$$y = 3e^{2x}.$$

9. Find $\frac{dy}{dx}$ of
$$2x + y^3 = cos(3y).$$

10. Find $\frac{dy}{dx}$ of
$$sin(y) + x^2 + 4y = cos(x).$$

11. Find
$$\int x^2 e^{3x} dx.$$

12. Find
$$\int x^2 ln|x| dx.$$

Chapter 2

First Order ODEs

In chapter one we looked briefly at the history of calculus and some interesting real world applications of differential equations. We also looked at key methods when differentiating and integrating. These will be used throughout the coming chapters.

Chapter two on the other hand is going to look at properties of differential equations and methods of solving first order differential equations. We will also look at graphical methods such as slope fields to predict what solutions look like.

In particular we will look at ordinary differential equations (ODEs). There are also partial differential equations (PDEs) which are differential equations that have partial derivatives terms. These are of particular interest in research but we will not look at these within this book. There is many books focused just on these and we suggest if the reader is interested they delve into the vast library out there.

2.1   Classification

Differential equations come in all shapes and sizes with a variety of properties. It is important to understand these properties as the method of solving a specific differential equation may differ depending on them. We will look at these classifications in the following section. In our further examples we will treat y as the dependent variable and x as the independent variable.

Until now we have used the notation Leibniz introduced. There are others way of expressing a derivative in differential equations. Another common way of representing a derivative is the dash notation introduced by Lagrange, and dot notation introduced by Newton. This is common with time derivatives in applied sciences.

$$\frac{d^2y}{dt^2} + \frac{dy}{dt} + y = 0$$

$$y'' + y' + y = 0$$

$$\ddot{y} + \dot{y} + y = 0$$

Above is the three different ways described above of writing the same equation.

2.1.1   Order

Definition: The order of a differential equation is the order of the highest derivative.

Let us look at some examples:

   1.

$$3\frac{d^3y}{dx^3} + x\frac{d^2y}{dx^2} = 5x^2 \tag{2.1}$$

2.

$$\frac{d^2y}{dx^2} - \frac{d^4y}{dx^4} = f(x) + \frac{dy}{dx} \qquad (2.2)$$

3.

$$\frac{y-x}{1+x^2} + \frac{dy}{dx} = 0 \qquad (2.3)$$

1. The first example has derivatives of order 3 and 2. Therefore the highest order is 3 so the equation is a third order differential.

2. The second example has derivative of order 2, 4 and 1 so we have a fourth order differential equation.

3. The final example has only one derivative and is order one.

Generally as the order of a differential equation increases the difficulty in solving it also increases. In chapter 2 we solve first order differential equations, in chapter two we solve second order differential equations, and in chapter 5 we solve higher order differential equations.

2.1.2 Linearity

Definition: A differential equation is linear if its first degree in the dependent variable.

The general form for the nth order differential is

$$a_n(x)\frac{d^ny}{dx^n} + a_{n-1}(x)\frac{d^{n-1}y}{dx^{n-1}} + \ldots + a_0(x)y = f(x) \qquad (2.4)$$

where $f(x)$ and a_k with k=0,1...n are functions of x and not of y.

If we look at our three examples used above. All three examples above are linear. An example of a non-linear equation is

$$(x + y^3)\frac{dy}{dx} + 5 = 0.$$

This equation is non-linear as the degree of the dependent variable y is multiplying the derivative of itself. Another example of a non-linear equation is

$$\frac{dy}{dx} = cos(y).$$

This is non-linear as the function on the right hand side of the equation is a function of y. An exception is y to the power of 1. This acts as a derivative of order 0. We see this in 2.4 where we have an $a_0(x)$ multiplying the y term.

Linear equations in comparison to non-linear equations are far easier to solve. Also their solutions have less complicated behaviours than their non-linear counterparts.

2.1.3 Homogeneity

The final property we will look at is whether or not the equation is homogeneous. If an equation is not homogeneous it is called non-homogeneous.

Definition: A differential equation is homogeneous if it has no term that is a function of the independent variable alone:

$$a_n(x)\frac{d^n y}{dx^n} + a_{n-1}(x)\frac{d^{n-1}y}{dx^{n-1}} + ... + a_0(x)y = 0. \qquad (2.5)$$

If again we look at the examples above; we can see that all three examples are non-homogeneous. An example of a homogeneous equation is

$$2\frac{d^2y}{dx^2} + x\frac{dy}{dx} - y = 0.$$

2.1.4 Initial Value and Boundary Value Conditions

When solving differential equations we often have to integrate meaning that we have constants of integration present.
Let

$$\frac{dy}{dx} = 2y.$$

We will learn how to solve this in later sections. The general solution for this differential equation is

$$y = Ae^{2x}$$

where A is a constant of integration. Depending how the system acts, A could be different values. We can plot the family of solutions for this to see how it acts and changes with different values of A. However to get a final solution to our problem we need a way of finding an exact value of A. To do this we would observe how the system acts at a given point and record this. Once we have this we would substitute these values into our solution to find a value for A. These as called the conditions of the equation and there are two main types we use, initial value and boundary conditions. Let use take an arbitrary second order differential equation.

$$\frac{d^2y}{dx^2} = f(x, y, y') \tag{2.6}$$

for $a \le x \le b$.

For this generic system we may have initial value conditions which are values for the solution and its first derivative at $x = a$ (often but necessarily $a = 0$), the initial values.

$$y(a) = A$$

$$y'(a) = B$$

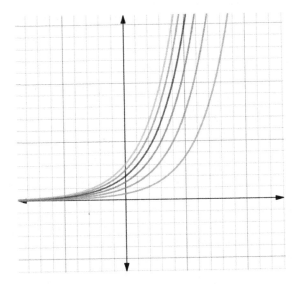

Figure 2.1: The graph of $y = Ae^{2x}$ for A=1,2,3,4,5,6.

The other set of conditions commonly given are boundary conditions. These are the solutions at two distinct points, usually the end points of the range.

$$y(a) = A$$

$$y(b) = B$$

If we consider the heat moving through a rod. This problem is solved with partial differential equations but it is a good way of visualising the conditions. Take a rod being heated at one end. The heat will move through the bar over time. If you are given values for the temperature and its derivative at a point in the bar, this is initial value conditions. If on the other hand you are given values for the temperature at either end of the bar these would be boundary conditions.

2.2 Slope Fields

Sometimes we want to find out what a solution looks like before we solve the equation. It is very powerful to know how a solution will act without the need to do the full calculations. Also we can learn about slope fields without any understanding of how to solve differential equations. Slope fields can be used with first order differential equations. We will look at how to plot these.

Example: Find the slope field for

$$\frac{dy}{dx} = \frac{1}{xy}.$$

Solution:

To be able to plot this slope field we look at coordinates and see what their slope looks like at those points. Let us take the point $(1,1)$. The value of $\frac{dy}{dx} = \frac{1}{(1)(1)} = 1$. This means at the coordinate $(1,1)$ the slope field has gradient 1. We would plot this on a pair of axis with a dash of gradient 1 at this point.

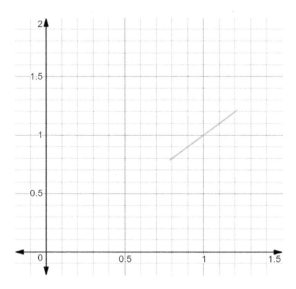

Figure 2.2: The slope field with the point at (1,1) plotted.

We repeat the process with multiple coordinates to get a feel of how the field acts. The more points we look at the more detailed the slope field plot will be.

x	y	$\frac{dy}{dx}$
1	1	1
-1	1	-1
-1	-1	1
1	2	$\frac{1}{2}$
2	1	$\frac{1}{2}$
2	2	$\frac{1}{4}$
0	1	Undefined

Plotting all the points from the table (and all the points in $x \in$

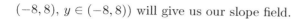

$(-8, 8)$, $y \in (-8, 8))$ will give us our slope field.

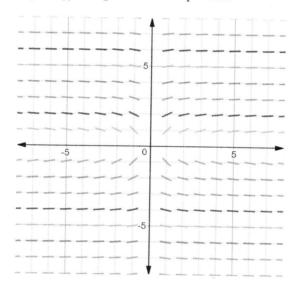

Figure 2.3: The complete slope field for $\frac{dy}{dx} = \frac{1}{xy}$.

We have to be careful using slope fields as they only give an indication of how the solution acts. Our case highlights this problem. The solution to our problem is

$$y = \sqrt{2} \cdot \sqrt{A - ln(x)}. \tag{2.7}$$

This solution only takes values for $x \in [0, \infty)$ for $A \neq 0$ however our slope field shows values over the whole of $\mathbb{R} \times \mathbb{R}$.

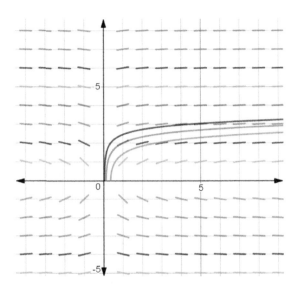

Figure 2.4: The complete slope field for $\frac{dy}{dx} = \frac{1}{xy}$ with graph of solutions for constant $A = 1, 2, 3$.

2.3   Finding Solutions to First Order Equations

Now we have an understanding of the properties of differential equations we are going to look at first order differential equations. In this section we will look at some key equations and methods to solve these.

The general form of a first order differential equation is

$$\frac{dy}{dx} + P(x)y = Q(x) \tag{2.8}$$

where $P(x)$ and $Q(x)$ are functions of x.

2.3.1 Basic Examples

In this section we will look at solving equations where P(x)=0. Note how the right hand side of the equation is only dependent on x.

$$\frac{dy}{dx} = Q(x) \qquad (2.9)$$

Let us take a first order linear ordinary differential equation.

Example: Solve

$$\frac{dy}{dx} = 9x^2$$

with the initial condition $y(1) = 18$.

Solution:

This is a nice equation and is fairly easy to solve. We can straight away integrate both sides of the equation with respect to x

$$\int \frac{dy}{dx} dx = \int 9x^2 dx$$

giving

$$y = 3x^3 + C$$

where C is the constant of integration. This gives us the general solution.

Definition: The general solution to a differential equation is the solution when the conditions, initial or boundary, are not taken into account.

To get the final solution we need to use initial conditions. The initial condition given to us was $y(1) = 18$. This means when we have $x = 1$ we get value of 18 for y. Substituting in our conditions give

$$18 = 3 + C$$

giving C=15 and therefore the final solution is

$$y = 3x^3 + 15.$$

This example was simple because integrating both sides worked however this will not always work for all first order differential equations.

2.3.2   Separation of Variables

The next type of equation we will look at is in the form

$$\frac{dy}{dx} = g(x)h(y). \tag{2.10}$$

Here we have a function of y multiplied by a function of x. We are going to start by dividing by $h(y)$ and then integrating both sides with respect to x.

$$\int \frac{dy}{dx} \frac{1}{h(y)} dx = \int g(x) dx$$

With the left hand side we need to use a change of variable. Taking $u = y(x)$ we get

$$du = y'(x)dx$$
$$= \frac{dy}{dx} dx.$$

Using these we obtain

$$\int \frac{1}{h(u)} du = \int g(x) dx. \tag{2.11}$$

We note is that the variable is just a name so 2.11 is the same as

$$\int \frac{1}{h(y)} dy = \int g(x) dx.$$

Let us look at a couple of examples. In these example we will go straight tot the integral form above.

Example: Find the general solution to the first order equation below

$$\frac{dy}{dx} = 6xy$$

Solution:

Taking all the y terms to the left hand side and the x terms to the right hand side

$$\int \frac{1}{y} dy = \int 6x \, dx$$

and integrating gives

$$ln(y) = 3x^2 + C.$$

We still need to rearrange to find a solution for y. Taking the inverse to the logarithm yields

$$y = e^{3x^2 + C}.$$

Currently we have the constant as a power of the exponential. If we want to find a value for this it is easier if it is not a power. By using laws of indices

$$y = e^{3x^2} e^C$$

where e^C is a constant. We can rename this A for convenience and obtain

$$y = Ae^{3x^2}.$$

This is our general solution. Applying initial conditions would give the final solution.

Example: Find the general solution to the first order equation below

$$\frac{dy}{dx} = -2$$

Solution:

Separating the variables and integrating

$$\int 1 dy = -\int 2 dx$$

giving

$$y = -2x + C.$$

Example: Find the general solution to the differential equation below

$$\frac{dy}{dx} = y^2$$

Solution:

Separating the variables and integrating gives

$$\int \frac{1}{y^2} dy = \int 1 dx.$$

Integrating both sides gives

$$-y^{-1} = x + C.$$

As we have y^{-1} we need to solve for y. By rearranging we obtain

$$y = -\frac{1}{x + C}$$

2.3.3 Integrating Factor

Our next first order equation is in the form

$$\frac{dy}{dx} + P(x)y = Q(x) \tag{2.12}$$

where $P(x)$ and $Q(x)$ are functions of x only. To solve this we use a method which involves an integrating factor. We first have to multiply the equation through by μ.

$$\mu\frac{dy}{dx} + \mu P(x)y = \mu Q(x)$$

Using the product rule

$$\frac{d}{dx}(\mu y) = \mu\frac{dy}{dx} + y\mu'$$

which gives

$$\mu P(x) = \mu'.$$

Thus we get a new equation involving μ and μ',

$$\mu\frac{dy}{dx} + \mu'y = \mu Q(x).$$

Once again we are going to use the product rule for the left hand side.

$$\frac{d}{dx}(\mu y) = \mu Q(x)$$

Integrating gives

$$\mu y + C = \int \mu Q(x)dx,$$

and solving for y,

$$y = \frac{\int \mu Q(x)dx - C}{\mu}.$$

Now that we have our final solution let us check out what μ actually is. Rearranging our equation of $P(x)$, μ and μ'

$$P(x) = \frac{\mu'}{\mu}.$$

and integrating

$$\int P(x)dx = \int \frac{\mu'}{\mu}dx$$

$$\int P(x)dx = log(\mu)$$

$$\mu = e^{\int P(x)}.$$

This is what we call our integrating factor. We can go directly from equation 2.12 to the solution by using our integrating factor.

$$y\mu = \int Q(x)\mu dx \qquad (2.13)$$

Now that we have seen the theory behind integrating factors let us see an examples.

Example: Solve

$$\frac{dy}{dx} + \frac{y}{x} = x^3$$

with initial condition of y(1)=2.

Solution:

We have that $P(x) = \frac{1}{x}$ and $Q(x) = x^3$. Using the formula above for the integrating factor we get

$$\mu = e^{\int \frac{1}{x}dx}$$

$$= e^{ln(x)}$$

$$= x$$

Substituting this into 2.13 we get

$$yx = \int x^3 \cdot x dx.$$

Integrating the right hand side

$$yx = \frac{x^5}{5} + C$$

and solving for y gives the general solution of

$$y = \frac{x^5}{5x} + \frac{C}{x}.$$

Using the initial condition $y(1) = 2$ we obtain the full solution

$$y = \frac{x^5 + 9}{5x}$$

Example: Solve

$$x\frac{dy}{dx} + y = 2x^3 e^{x^2}$$

with initial condition $y(1) = e$.

Solution:

First we need to put the equation in the form where we can use an integrating factor. Dividing through by x we get

$$\frac{dy}{dx} - \frac{y}{x} = 2x^2 e^{x^2}.$$

In this case $P(x) = \frac{1}{x}$ and $Q(x) = 2x^2 e^{x^2}$. We get the same integrating factor as the previous example. This then gives

$$yx = \int 2x^2 e^{x^2} \cdot x dx.$$

To find the value of the integral we need to use a substitution of $u = x^2$.

$$yx = \int ue^u du$$

For the next step we need to integrate by parts

$$yx = ue^u - e^u.$$

Substituting back gives

$$yx = x^2 e^{x^2} - e^{x^2} + C$$

and by solving for y

$$y = xe^{x^2} - \frac{e^{x^2}}{x} + \frac{C}{x}.$$

Using our initial condition of $y(1) = e$ we see that

$$e = C.$$

Giving our final solution as

$$y = xe^{x^2} - x^{-1}(e^{x^2} - e).$$

Plotting this gives an interesting graph with an asymptote at $x = 0$ (remember an asymptote is a value in which the graph cannot cross).

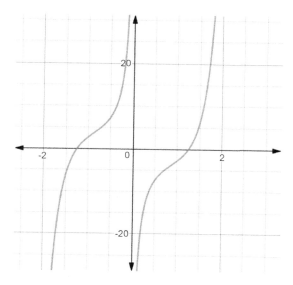

Figure 2.5: Plot of $y = xe^{x^2} - x^{-1}(e^{x^2} + e)$.

2.3.4 Substitution

Sometimes we have an equation that may be too complicated to solve in its current form meaning we have to change the equation to be able to solve it. What we can do is use a substitution to make the equation easier to deal with. This can be particularly useful when we have a non-linear equation, as a good choice of substitution can convert this into a linear equation which can be solved using the methods above.

If we look at the following non-linear equation we can use a substitution to help us solve this.

Example: Find the general solution to the first order equation below using the substitution of $z = \frac{y}{x}$.

$$x^2 \frac{dy}{dx} = 6y^2 + xy$$

Solution:

First let us plot the slope field to see how the solution will act.

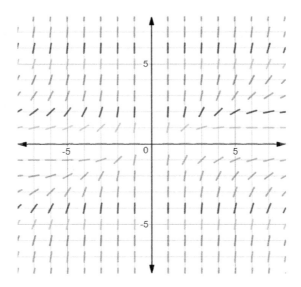

Figure 2.6: The slope field for $x^2 \frac{dy}{dx} = 6y^2 + xy$

We can see this is not a linear equation as the dependent variable y has a power other than 1 (even when manipulated). If we look at the equation we can get it into a form where we substitute z in directly. This may not always be the case but for this example we can. Dividing through by x^2 we see

$$\frac{dy}{dx} = \frac{6y^2 + xy}{x^2}.$$

Splitting the fraction into two fractions we see

$$\frac{dy}{dx} = \frac{6y^2}{x^2} + \frac{y}{x}.$$

We have a derivative of y with respect to x. To replace this with a derivative we need to use the product rule on $y = xz$ (the rearranged form of our substitution). This gives

$$\frac{dy}{dx} = x\frac{dz}{dx} + z.$$

Substituting this and $z = \frac{y}{x}$ we obtain

$$x\frac{dz}{dx} + z = 6z^2 + z.$$

The z terms cancel leaving us with an equation that we can use separation of variables on.

$$6\int \frac{1}{z^2}dz = \int \frac{1}{x}dx$$

Integrating and rearranging gives our general solution for z

$$z = -\frac{6}{ln(x) + A}.$$

We note that this is not the general solution we desire. Our dependent variable is currently z when we were working with y originally. Therefore we substitute back and gain our general solution in y,

$$y = -\frac{6x}{ln(x) + A}.$$

2.3.5 Bernoulli Equation

Bernoulli was a Swiss mathematician of the 18th century. He is particularly remembered for his work in fluid mechanics. One of his most famous principles, which was named after him, gives rise to a differential equation form we are interested in.

Theorem 2: (Bernoulli's Principle) Points of greater fluid speed will have less pressure than points of lower fluid speed for horizontal flow.

This is a statement for the conservation of energy for fluid mechanics produces differential equations of the following form

$$\frac{dy}{dx} + P(x)y = Q(x)y^n \tag{2.14}$$

sometimes written as

$$y^{-n}\frac{dy}{dx} + P(x)y^{1-n} = Q(x). \tag{2.15}$$

To solve Bernoulli equations we are going to use the method substitution we looked at in the previous section. As this is a specific case we have a substitution that is always the same. This substitution is $z = y^{1-n}$. We also need to know the substitution for the derivative. If we take $z = y^{1-n}$ and take the derivative with respect to x we see

$$\frac{dz}{dx} = \frac{d}{dx}y^{1-n}.$$

This is not an instant result and we have to use the implicit differentiation we looked at in chapter 1. So by using implicit differentiation on the right hand side

$$\frac{dz}{dx} = \frac{d}{dy}y^{1-n}\frac{dy}{dx} = (1-n)y^{-n}\frac{dy}{dx}.$$

Substituting this into equation 2.14 we obtain the new equation

$$\frac{1}{1-n} \cdot \frac{dz}{dx} + zP(x) = Q(x).$$

As n is not dependent on z we can use an integrating factor to solve this equation. The value for n is often given and is a real value.

Example: Find the general solution to the Bernoulli equation below using a substitution method.

$$\frac{dy}{dx} - \frac{1}{x}y = -y^4$$

Solution:

Here we have a Bernoulli equation where $n = 4$. This then means that we should use the substitution $z = y^{-3}$ giving the derivative to be

$$\frac{dz}{dx} = -3y^{-4}\frac{dy}{dx}$$

using implicit differentiation. Substituting this gives the linear first order equation

$$-\frac{1}{3} \cdot \frac{dz}{dx} - \frac{1}{x}z = -1.$$

Rearranging gives

$$\frac{dz}{dx} + \frac{3}{x}z = 3.$$

As now we have a differential equation of the form $y' + yP(x) = Q(x)$ we can use an integrating factor. Using the integrating factor of $\mu = e^{\int \frac{3}{x}dx} = x^3$ gives

$$x^3z = \int (3)(x^3)dx$$

$$= \frac{3}{4}x^4 + A$$

Rearranging and substituting y back in

$$y^{-3} = \frac{3x^4 + 4A}{4x^3}.$$

Therefore our general solution is

$$y = \sqrt[3]{\frac{4x^3}{3x^4 + 4A}}$$

We need to be careful when quoting this as we need three solutions. Our three solutions will be

1.

$$y = \frac{\sqrt[3]{4}x}{\sqrt{3x^4 + 4A}}$$

2.

$$y = \frac{\sqrt[3]{-4}x}{\sqrt{3x^4 + 4A}}$$

3.

$$y = -\frac{\sqrt[3]{-1}\sqrt[3]{4}x}{\sqrt{3x^4 + 4A}}.$$

2.4 Singularities

When finding a general solution to a differential equation we often use integration within our calculations. Whenever we use integration we end up with constants of integration. With the use of initial and boundary conditions these constants take on a value. However some conditions may cause the solution to 'blow up' and become undefined.

Let us take a first order differential equation,

$$\frac{dy}{dx} = y^2,$$

with $y(0) = A$.

The general solution to this equation is

$$y = \frac{A}{1 - Ax}.$$

If $x = 1/A$ then the solution will become undefined. As the initial condition A changes, the point at which the equation becomes undefined changes. These are called movable singularities.

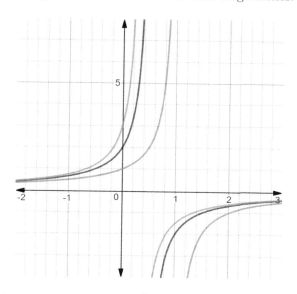

Figure 2.7: Graph for $\frac{A}{1-Ax}$ for values of A=1, 2, 3.

Another type of singularity is a fixed singularity. These are singularities that do not change as the conditions change.

Let us take the first order differential equation

$$\frac{dy}{dx} + \frac{y}{x} = x^2.$$

The solution to the equation above is

$$y = \frac{x^3}{4} + \frac{A}{x}.$$

We can see from the equation above that the solution 'blows up' when $x = 0$. This singularity does not depend on the conditions. We call this a fixed singularity.

2.5 Uniqueness Theorem

Theorem 3: For an arbitrary second order differential equation in the form

$$\frac{dy}{dx} + a_2(x)y = a_3(x) \tag{2.16}$$

with the initial values $y(x_0) = y_0$.

If $a_1(x)$ and $a_2(x)$ are continuous function over an interval $x \in [\alpha, \beta]$ containing $x = x_0$ then there exists a unique solution and it exists over the interval.

Let us look at an example.

Example: Find the largest interval where $x_0 = 1$ of

$$\frac{dy}{dx} + \frac{y}{x+1} = \frac{1}{sin(x)}.$$

Solution:
There are discontinuities at $x = -1, 0, \pm n\pi$. The largest interval is

$(0, \pi)$.

Example: Is there a solution to the linear differential equation

$$\frac{dy}{dx} + \frac{y}{x-2} = e^x.$$

y(0)=2

Solution:
The differential equation above has $a_1(x) = \frac{1}{x-2}$ and $a_2(x) = e^x$. $a_2(x)$ is continuous everywhere but $a_1(x)$ is discontinuous at $x = 2$. As our initial value $y(0) = 2$ we see that there is no solution.

Example: Is there a solution to the linear differential equation

$$x\frac{dy}{dx} + y = cos(x).$$

y(0)=1

Solution:
To be able to use the theorem above we first need to get rid of the coefficient of the dy/dx term.

$$\frac{dy}{dx} + \frac{y}{x} = \frac{cos(x)}{x}$$

We now note that both $a_1(x)$ and $a_2(x)$ have discontinuities at $x = 0$ but continuous at all other values of x. As our initial condition is $y(0) = 1$ we see that this differential equation has a solution.

2.6 Real World Applications I

Now that we have seen how to solve first order ODEs let us now apply these skills to some real world examples.

Example: Ryan has £1000 to invest. He finds a saving account that offers 5% continuous compound invests. Model this and find out how long it will take for his money to double.

Solution:

The initial amount M, the rate of change of money with respect to time $\frac{dM}{dt}$, and interest rate $r = 0.05\%$ are given. Putting these together

$$\frac{dM}{dt} = 0.05M$$

with the initial amount $M = 1000$. This is a separable equation like the ones solved earlier. First we take all the terms of M to the left hand side and the terms of t to the right hand side.

$$\int \frac{1}{M} dM = \int 0.05 dt$$

Then integrating we get

$$ln(M) = 0.05t + A.$$

Solving for M gives us the general solution

$$M = Ae^{0.05t}.$$

Using our initial conditions M(0)=1000 our final solution is

$$M = 1000e^{0.05t}.$$

To find the point at which the money doubles we substitute $M = 2000$ and solve. Doing this gives

$$t = 20ln(2) \approx 14.$$

Example: A lake is being set up to be used as a site for sport fishing. The owner initially puts 1000 carp into the lake. It is observed that each year the carp population has a growth rate coefficient of 0.3. It is believe that the lake has not got the resources for more than 2000 fish. Model the rate of the carp increase over time.

Solution:

This problem requires us to use the logistic model by Verhulst. Using the information above we obtain

$$\frac{dy}{dx} = 0.3\left(1 - \frac{y}{2000}\right)y.$$

This is in fact a Bernoulli equation. Expanding the brackets and diving through by y^2 gives the usual form

$$y^{-2}\frac{dy}{dx} = 0.3y^{-1} - \frac{0.3}{2000}.$$

Using a substitution $z = y^{-1}$ and $\frac{dz}{dx} = -y^{-2}\frac{dy}{dx}$. Substituting these in we get

$$\frac{dz}{dx} + 0.3z = \frac{0.3}{2000}.$$

We can solve this by using an integrating factor $\mu = e^{-0.3x}$,

$$ze^{0.3x} = -\frac{0.3}{2000}\int e^{0.3x}dx$$

giving

$$z = \frac{1}{2000} + Be^{-0.3x} = \frac{2000e^{0.3x}}{e^{0.3x} + 2000B}.$$

Substituting y back into the equation and rearranging gives

$$y = \frac{2000e^{0.3x}}{e^{0.3x} + 2000B}.$$

Using our initial condition we see that $B = \frac{1}{2000}$. Our final solution
is

$$y = \frac{2000e^{0.3x}}{e^{0.3x} + 1}.$$

We can plot this. We can see how the graph tends to 2000 but
never crosses it. This shows how the population grows over time.
The growth starts quickly as the environment has plenty of resources
but slows as it approaches the limit of these. We can see from the
graph that at the 3 year mark it has gained nearly 500 carp but
takes almost to the 20 year mark to reach near its capacity.

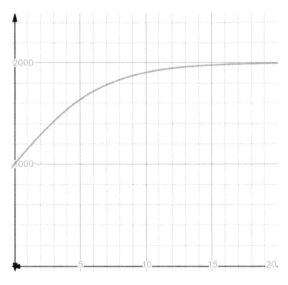

Figure 2.8: Graph for $y = \frac{2000e^{0.3x}}{e^{0.3x}+1}$.

2.7 Exercises

1. State whether the following problem is linear/non-linear, initial value/boundary value conditions, and state the order.

   i)
   $$\frac{dy}{dx} + x = sin(x)$$
   with y(0)=0.

   ii)
   $$y^2\frac{dy}{dx} + 3y = 4x$$
   with y(0)=2.

   iii)
   $$\frac{d^3y}{dx^3} - y = 2x$$
   with y(0)=y'(0)=y"(0)=0.

   iv)
   $$\frac{d^2y}{dx^2} - \frac{dy}{dx} = cos(y)$$
   with y(0)=3, y(5)=10.

2. State whether the following are true or false

   i)
   $$\left(\frac{d^3y}{dx^3}\right)^4 - \frac{dy}{dx} = cos(x)$$
   is a forth order ODE.

   ii)
   $$\frac{dy}{dx} - 3y = e^{2y}$$
   is a linear equation.

iii)

$$\frac{d^2y}{dx^2} = 4y$$

is homogeneous.

3. Find the general solution to

$$\frac{dy}{dx} + x^2 = 4x$$

4. Find the general solution to

$$4\frac{dy}{dx} + 6y = 0.$$

5. Solve

$$\frac{dy}{dx} - 5y = 0$$

with initial conditions y(0)=2.

6. Solve

$$\frac{dy}{dx} = 3x^2y$$

with initial conditions y(0)=4.

7. Solve

$$\frac{dy}{dx} - \cos(x)y = \cos(x)$$

with the initial condition y(0)=1

8. Find the general solution to $\frac{dy}{dx} + \frac{y}{x} = \cos(x)$.

9. Find the general solution to the Bernoulli equation

$$\frac{dy}{dx} + \frac{1}{x}y = y^2$$

with the transformation $z = y^{1-n}$.

10. Find the general solution to

$$\frac{dy}{dx} + y = 6e^{2x}$$

11. Find the general solution to

$$x\frac{dy}{dx} - 2y = x^3 \sin(x)$$

Chapter 3

Second Order ODEs

In chapter 2 we looked at first order equations and some of the methods to solve these. In this chapter we are going to move onto second order equations. These equation appear often in mechanical systems as well as chemical processes. We are going to look in detail at equations with constant coefficients and then venture out to look at a form of equation (Euler equation) which has varying constants.

3.1 Constant Coefficients

For this section we are going to look at constant coefficients. These are equations of the form

$$a\frac{d^2y}{dx^2} + b\frac{dy}{dx} + cy = f(x), \tag{3.1}$$

with the homogeneous form

$$a\frac{d^2y}{dx^2} + b\frac{dy}{dx} + cy = 0. \tag{3.2}$$

We are going to start by looking at the homogeneous equation. We begin by assuming a solution to the equation is $y = e^{mx}$. Taking this as a solution, we substitute it into 3.2.

$$am^2 e^{mx} + bm e^{mx} + c e^{mx} = 0$$

If we factor out e^{mx} we have a quadratic in m.

$$e^{mx}(am^2 + bm + c) = 0$$

We notice that am exponential is never zero so the quadratic must be equal to zero. We can solve this using the quadratic formula giving our general solution.

$$am^2 + bm + c = 0 \tag{3.3}$$

3.3 is known as the characteristic equation , sometimes called auxiliary equation.

Depending on the outcome of the quadratic formula the general solution for the ODE is different.

3.1.1   Unique Real Roots

If we take our solutions to the auxiliary equation and we get two real, distinct roots m (this happens when the discriminant is strictly greater than 0) giving two solutions being $y_1 = Ae^{m_1 x}$ and $y_2 = Be^{m_2 x}$. Our general solution is built from these.

Theorem 4: (Principle of Superposition) If $y_1(x)$ and $y_2(x)$ are two solutions to a linear homogeneous differential equation then the sum of the solutions is also a solution $y = y_1(x) + y_2(x)$

We use the principle of superposition to state that the solution for our differential equations are

$$y = Ay_1 + By_2 + ... + \gamma y_n.$$

For our example (unique real roots) we get

$$y = Ae^{m_1 x} + Be^{m_2 x}. \tag{3.4}$$

Let us now look at some examples.

Example: Solve

$$\frac{d^2y}{dx^2} + 5\frac{dy}{dx} + 4y = 0$$

with $y(0) = 0$ and $y'(0) = 3$.

Solution:

We start by using the auxiliary equation

$$m^2 + 5m + 4 = 0.$$

We can find m either by using the quadratic equation or by factorising. This gives us real values, $m = -4$ and $m = -1$. Substituting these into 3.4 gives

$$y = Ae^{-x} + Be^{-4x}.$$

This is our general solution. To find our full solution we will use our initial conditions. What we note is that for our initial conditions we need the derivative of the general solution,

$$\frac{dy}{dx} = -Ae^{-x} - 4Be^{-4x}.$$

Now when we substitute our initial conditions into these we get two simultaneous equations. For this example we get;

$$0 = A + B$$
$$3 = -A - 4B.$$

Solving this pair of simultaneous equations we obtain $A = 1$ and $B = -1$ giving the final solution as

$$y = e^{-x} - e^{-4x}$$

Example: Find the general solution to the first order equation below

$$\frac{d^2y}{dx^2} - 4y = 0$$

Solution:

This second order equation has no first order derivative. We follow the same steps as our first example however our value for b in the quadratic equation as zero. We are going to find just the general solution for this equation. First we use the auxiliary equation.

$$m^2 - 4 = 0$$

Remember that when square-rooting both sides we get a positive and negative value. This gives us our two real distinct roots. From above we get $m = \pm 2$.

$$y = Ae^{2x} + Be^{-2x}$$

3.1.2   Complex Roots

Above we had that the discriminant was strictly greater than zero. This section will focus on when the discriminant is strictly less than zero, ie we have complex roots

$$m = a \pm ib.$$

Substituting into equation 3.4 we get

$$y = c_1 e^{(a+bi)x} + c_2 e^{(a-bi)x}.$$

Note that we are using c_1 and c_2 as my constants for now as later on we will use A and B. Expanding the brackets and using laws of indices

$$y = c_1 e^{ax} e^{bix} + c_2 e^{ax} e^{-bix}.$$

Both terms on the right hand side has e^{ax} as a factor. We can factorise this out

$$y = e^{ax}(c_1 e^{bix} + c_2 e^{-bix}).$$

Now we are going to use Euler's formula to replace the exponentials with imaginary powers.

$$e^{i\theta} = cos(\theta) + isin(\theta)$$

By using this and the fact that $sin(x)$ is an odd function such that $sin(-\theta) = -sin(\theta)$ we obtain

$$y = e^{ax}(c_1(cos(bx) + isin(bx)) + c_2(cos(bx) - isin(bx))$$

If we group together the terms with cosine and sine respectively

$$y = e^{ax}((c_1 + c_2)cos(bx) + (c_1 - c_2)isin(bx)).$$

As we have not used A and B as constants we can call $c_1 + c_2{=}A$ and $(c_1 - c_2)i{=}B$ we obtain the general solution

$$y = Ae^{ax}cos(bx) + Be^{ax}sin(bx) \qquad (3.5)$$

Let us look at some examples.

Example: Solve

$$\frac{d^2 y}{dx^2} + 2\frac{dy}{dx} + 5y = 0$$

with $y(0) = 0$ and $y'(0) = 4$.

Solution:

Using the auxiliary equations and the quadratic formula we obtain

$m = -1 \pm 2i$. Substituting these into 3.5 gives the general solution as

$$y = Ae^{-x}cos(2x) + Be^{-x}sin(2x).$$

We see that before we can use the initial conditions we need to differentiate our general solution. From this we obtain

$$\frac{dy}{dx} = -2Asin(2x) + 2Bcos(2x).$$

Now we have this we can use the initial conditions.

$$0 = A$$
$$4 = 2B$$

This gives $A = 0$ and $B = 2$. Our final solution is therefore

$$y = 2e^{-x}sin(2x).$$

Example: Solve

$$\frac{d^2y}{dx^2} + 9y = 0$$

with y(0)=0 and y'(0)=9.

Solution:

Like the second example in the unique real roots section this equation has no first order term. Using the auxiliary equation we obtain $m = \pm 3i$. Notice how we have no real part and so the exponential in our general solution will equal one. This then gives

$$y = Acos(3x) + Bsin(3x).$$

Once again we have to differentiate our general solution to be able to use the initial conditions.

$$\frac{dy}{dx} = -3Asin(3x) + 3Bcos(3x).$$

Now we have this we can use the initial conditions.

$$0 = A$$
$$9 = 3B$$

This gives $A = 0$ and $B = 3$. Our final solution is therefore

$$y = 3sin(3x).$$

3.1.3 Repeated Roots

So far we have looked at when the discriminant is strictly positive and strictly negative. Now we are going to look at what happens when we have the discriminant equal to zero, ie. repeated roots to the auxiliary equation. Let $m = a$. This gives

$$y = Ae^{ax} + Be^{ax} = Ce^{ax}$$

But as we have two initial conditions it will find two different values for C which cannot happen. We now have to look at another possible general solution. We will assume the solution is a function of x multiplied by the exponential.

$$y = f(x)e^{nx}$$

However we know that our discriminant equals zero as we have repeated roots. This means that in this case our n we used above is actually $-\frac{b}{2a}$. We have

$$y = f(x)e^{-\frac{bx}{2a}}$$
$$\frac{dy}{dx} = f'(x)e^{-\frac{bx}{2a}} - \frac{b}{2a}f(x)e^{-\frac{bx}{2a}}$$
$$\frac{d^2y}{dx^2} = f''(x)e^{-\frac{bx}{2a}} - \frac{b}{2a}f'(x)e^{-\frac{bx}{2a}} - \frac{b}{2a}f''(x)e^{-\frac{bx}{2a}} + \frac{b^2}{2a}f'(x)e^{-\frac{bx}{2a}}$$

With the above we substitute into 3.2 and simplify

$$e^{-\frac{bx}{2a}}\left(af''(x) - \frac{1}{4a}(b^2 - 4ac)f(x)\right) = 0.$$

Again we use the fact that the discriminant is zero.

$$ae^{-\frac{bx}{2a}}f''(x) = 0$$

As an exponential can not be zero means that either $a = 0$ or $f''(x) = 0$. If $a = 0$ then we get a trivial solution so $f''(x)$ must equal zero.

$$f''(x) = 0$$
$$f'(x) = A$$
$$f(x) = Ax + B$$

With f(x) we can find our general solution.

$$y = e^{nx}(Ax + B)$$

or

$$y = Axe^{nx} + Be^{nx} \qquad\qquad (3.6)$$

Now let us look at some examples.

Example: Solve

$$\frac{d^2y}{dx^2} - 4\frac{dy}{dx} + 4y = 0$$

with $y(0) = 2$ and $y'(1) = e^2$.

Solution:

Using the auxiliary equation we get $(m - 2)^2 = 0$ giving $m = 2$. Substituting this into 3.6 we get out general solution.

$$y = Axe^{2x} + Be^{2x}$$

Taking the derivative for use with the initial conditions, making note we have to differentiate by parts on the first term, gives

$$\frac{dy}{dx} = Ae^{2x} + 2Axe^{2x} + 2Be^{2x}.$$

Using the initial conditions we have the pair of simultaneous equations

$$2 = B$$
$$e^2 = 3Ae^2 - 4e^2$$

obtain $A = -1$ and $B = 2$ with the full solution

$$y = -xe^{2x} + 2e^{2x}.$$

We are now going to look at a boundary value problem.

Example: Solve

$$\frac{d^2y}{dx^2} + 6\frac{dy}{dx} + 9y = 0$$

with $y(0) = 2$ and $y(5) = 6e^{-15}$.

Solution:

Using the auxiliary equation we get $m = -3$ gives the general solution to be

$$y = Axe^{-3x} + Be^{-3x}.$$

So for our simultaneous equations we get

$$2 = B$$
$$6e^{-15} = 5Ae^{-15} + 2e^{-15}$$

giving $A = \frac{4}{5}$ and $B = 2$. Our final solution is

$$y = \frac{4}{5}xe^{-3x} + 2e^{-3x}.$$

3.1.4   Non-Homogeneous Equations

In the previous parts of this chapter we looked at homogeneous second order ODEs which took the form

$$a\frac{d^2y}{dx^2} + b\frac{dy}{dx} + cy = 0.$$

We note that here we took $f(x) = 0$ for our generic case. Now we are going to look non-homogeneous second order ODEs, ie when $f(x) \neq 0$.

When working with a non-homogeneous equation we start by solving the equation as if it was homogeneous ignoring $f(x)$. When we have solve this we call it the complementary function. After we have obtained this we need to find the particular solution. To obtain the particular solution we need to look at $f(x)$. We make a 'guess' for the solution based upon what our $f(x)$ is. We look for a function of the same form as this. Substituting this back in to the original equation will find the constants.

Let us now look at an example.

Example: Solve

$$\frac{d^2y}{dx^2} + 16y = e^{2x}$$

with initial conditions $y(0) = 0$ and $y'(0) = 0$.

Solution:

First we treat the equation as if it was homogeneous by ignoring the e^{2x}.

$$\frac{d^2y}{dx^2} + 16y = 0.$$

Using the auxiliary equation we obtain $m = \pm i4$. As these are complex roots we use the general solution of harmonics

$$y_c = A cos(4x) + B sin(4x).$$

Note we are using y_c as the solution for the complementary function. Now we have this we need to find the particular solution. As the $f(x) = e^{2x}$ our 'guess' needs to follow the same form. We keep the same power of the exponential and multiply by a constant. Our 'guess' will be $y = Ce^{2x}$. To be able to substitute this into our original equation we need to find $\frac{d^2y}{dx^2}$.

$$y = Ce^{2x}$$

$$\frac{dy}{dx} = 2Ce^{2x}$$

$$\frac{d^2y}{dx^2} = 4Ce^{2x}$$

Now that we these we can substitute them back into our differential equation and solve for C obtaining

$$4Ce^{2x} + 16Ce^{2x} = e^{2x}.$$

Dividing through by e^{2x} and solving for C we obtain $C = \frac{1}{20}$. Thus we obtain for our particular solution

$$y_p = \frac{1}{20}e^{2x}.$$

By putting together the complementary and particular solutions we obtain the general solution

$$y = A cos(4x) + B sin(4x) + \frac{1}{20}e^{2x}.$$

To get the final solution we need to use our initial conditions. Before we can use these we need to find the derivative of the general solution

$$\frac{dy}{dx} = -4A sin(4x) + 4B cos(4x) + \frac{1}{10}e^{2x}.$$

Using $y(0) = 0$ we see that

$$0 = A + \frac{1}{20}$$

giving $A = -\frac{1}{20}$. Using $y'(0) = 0$ we see that

$$0 = 4B + \frac{1}{10}$$

giving $b = -\frac{1}{40}$. This example was slightly easier than most with regards to conditions as it did not give simultaneous equations with A and B popping straight out. Therefore our final solution is

$$y = -\frac{1}{20}cos(4x) - \frac{1}{40}sin(4x) + \frac{1}{20}e^{2x}.$$

We have plot the solution below in figure 3.1. From this we can look at the behaviour of our solution. In the region of $(-\infty, 0)$ the solution gives harmonic motion oscillating above and below the axis. This is due to the exponential function having very little effect in this region. However when we move into positive values of x we see the exponential taking a lot more effect and the solution increasing rapidly.

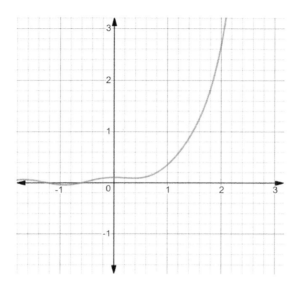

Figure 3.1: Graph for $y = \frac{1}{20}cos(4x) - \frac{1}{40}sin(4x) + \frac{1}{20}e^{2x}$

Example: Solve

$$\frac{d^2y}{dx^2} + 6\frac{dy}{dx} + 9y = cos(2x)$$

with initial conditions $y(0) = 0$ and $y'(\pi) = 0$.

Solution:

Once again we look at the homogeneous 'part' of the equation first. Using the auxiliary equation on this we find that we have repeated roots $m = -3$. So the complementary function is

$$y_c = Ae^{-3x} + Bxe^{-3x}$$

Now we look to find the particular solution. As our function of x is $cos(2x)$ a 'guess' in the same form needs the same coefficient of x. As we have the cosine function we will include this however we have to be careful. Whenever we have sine or cosine as part of the $f(x)$ we need to include both harmonics in our 'guess'; $y = C cos(2x) + D sin(2x)$. Working out the derivatives

$$y = C cos(2x) + D sin(2x)$$

$$\frac{dy}{dx} = -2C sin(2x) + 2D cos(2x)$$

$$\frac{d^2y}{dy^2} = -4C cos(2x) - 4D sin(2x)$$

Substituting these into our original equation we obtain

$$5C cos(2x) + 5 sin(2x) - 12C sin(2x) + 12D cos(2x) = cos(2x)$$

We need to find the values of C and D. This may appear be a problem as we only have one equation. We can overcome this by doing a trick called equating coefficients. For this we look at all the sine terms and cosine terms separately and these will need equality either side of the equals sign. This gives us a pair of simultaneous equations

$$cos(2x)[5C + 12D] = 1$$

$$sin(2x)[5D - 12C] = 0$$

Solving these gives $C = \frac{5}{169}$ and $D = \frac{12}{169}$ meaning that the particular solution is

$$y_p = \frac{5}{169} cos(2x) + \frac{12}{169} sin(2x).$$

So our general solution is

$$y = A e^{-3x} + B x e^{-3x} + \frac{5}{169} cos(2x) + \frac{12}{169} sin(2x).$$

Before we can use the initial conditions we need the derivative of our general solution.

$$\frac{dy}{dx} = -3A e^{-3x} - 3B e^{-3x} - \frac{10}{169} sin(2x) + \frac{24}{169} sin(2x).$$

Using our initial conditions

$$y(0) = 0 = A + \frac{5}{169}$$

giving $A = \frac{5}{169}$, and

$$y'(\pi) = 0 = \frac{5}{169}e^{-3\pi} + B\pi e^{-3\pi} + \frac{5}{169}$$

giving $B = -\frac{5}{169}[\frac{1+e^{3\pi}}{\pi}]$. Once again we were lucky with our conditions and A and B popped out nicely without further calculations.

Putting all this together our final solution becomes

$$y = \frac{5}{169}e^{-3x} - \frac{5}{169}\left[\frac{1+e^{3\pi}}{\pi}\right]xe^{-3x} + \frac{5}{169}cos(2x) + \frac{12}{169}sin(2x)$$

Below is a table of 'guesses' to make for the particular solution.

Function of x	Guess
x	Ax
x^2	$Ax^2 + Bx$
x^n	$Ax^n + Bx^{n-1} + ...$
$cos(ax)$	$Acos(ax) + Bsin(ax)$
$sin(ax)$	$Acos(ax) + Bsin(ax)$
$cos(ax) + sin(ax)$	$Acos(ax) + Bsin(ax)$
be^{ax}	Ae^{ax}

3.2 Euler Equation

So far we have only looked at second order differential equations with constant coefficients. Now we are going to look at a special case where the coefficient of $\frac{d^2y}{dx^2}$ is x^2 and the coefficient of $\frac{dy}{dx}$ is x. We call this the Euler equation;

$$ax^2\frac{d^2y}{dx^2} + bx\frac{dy}{dx} + cy = 0. \tag{3.7}$$

The appearance of the x^2 and x mean the methods looked at previously do not work with the equation as it is. However by transforming with a new independent variable t, such that $x = e^t$, we see that the equation becomes a constant coefficient problem.

With this new independent variable we see that the relation between first order derivative in x and t is

$$\frac{dy}{dx} = \frac{dy}{dt}\frac{dt}{dx} = \frac{1}{x}\frac{dy}{dt}$$

which is

$$x\frac{dy}{dx} = \frac{dy}{dt}.$$

For the second order derivatives

$$\frac{d}{dx}\left(\frac{dy}{dx}\right) = \frac{d}{dx}\left(\frac{1}{x}\frac{dy}{dx}\right) = -\frac{1}{x^2}\left(\frac{dy}{dt}\right) + \frac{1}{x^2}\frac{d^2y}{dx^2}$$

which is

$$x^2\frac{d^2y}{dx^2} = \frac{d^2y}{dt^2} - \frac{dy}{dt}.$$

Using these we can turn the Euler equation into a constant coefficient problem.

$$a\frac{d^2y}{dx^2} + (b-a)\frac{dy}{dx} + cy = 0 \tag{3.8}$$

Let us now look at an example.

Example: Solve

$$x^2\frac{d^2y}{dx^2} + 2x\frac{dy}{dx} - 6y = 0.$$

with $y(1) = 1$ and $y'(1) = 0$.

Solution:

Using the transformation in 3.8 with $a = 1$, $b = 2$ and $c = -6$ we obtain the constant coefficient problem

$$\frac{d^2y}{dx^2} + \frac{dy}{dx} - 6y = 0.$$

Now that we have a second order ODE with constant coefficients we can use the methods learnt previously in this chapter to solve this. Using the auxiliary equation we see that $m = -3$ or $m = 2$. As these are real distinct roots this means our general solution is

$$y = Ae^{2t} + Be^{-3t}.$$

As our initial conditions are with respect to x we must first transform back from a variable of t to x,

$$y = Ax^2 + Bx^{-3}.$$

We can now use the initial conditions and obtain the pair of simultaneous equations

$$1 = A + B$$
$$0 = 2A - 3B.$$

Solving these get a value for our constants, $A = \frac{3}{5}$ and $B = \frac{2}{5}$. Our final solution is

$$y = \frac{3}{5}x^2 + \frac{2}{5}x^{-3}.$$

It is interesting to note that if $x = 0$ we get a singularity. This is a fixed singularity as it is not dependent on the values of the constants.

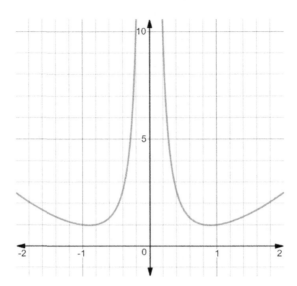

Figure 3.2: The graph of $y = \frac{3}{5}x^2 + \frac{2}{5}x^{-3}$ showing a singularity at $x = 0$.

3.3   Reduction of Order

Sometimes it is beneficial to reduce the order of our equation to make it easier to solve. In this section we are going to look at the method of reducing the order of a second order differential equation to a first order problem and then solving it. We are going to look at an example that we are already able to solve using methods from previous sections of this chapter but this will hopefully help with solving more advanced problems later.

To be able to use this method we need to already have one solution, $y_1 = f(x)$. In real life this may be an observation but we will always give it in any questions we have. Using this solution we then assume the second solution is $y_2 = u(x)f(x)$. Taking the derivatives we then can plug these into our original equation and simplify. It just so happens that using this method cancels all term of $u(x)$ with degree 1. Now we have a differential of second and first order derivatives. Letting $w = u'$ and so $w' = u''$ we can reduce the order and then solve more simply.

Example: Solve

$$\frac{d^2y}{dx^2} + \frac{dy}{dx} - 6y = 0$$

with solution of $y_1 = e^{2x}$.

Solution:

We first assume $y_2 = e^{2x}u(x)$. Then

$$y_2' = u'e^{2x} + 2ue^{2x},$$
$$y_2'' = u''e^{2x} + 4u'e^{2x} + 4ue^{2x}.$$

Substituting these in our original equation we see

$$u''e^{2x} + 4u'e^{2x} + 4ue^{2x} + u'e^{2x} + 2ue^{2x} - 6ve^{2x}.$$

Simplifying this gives

$$u''e^{2x} + 5u'e^{2x} = 0.$$

Now we let $w = u'$ and $w' = u''$. Substituting this and canceling the exponentials we obtain

$$w' + 5w = 0.$$

This is a first order equation and can be solved with separation of variables giving

$$w = Ae^{-5x}.$$

Substituting back in u' we get another first order equation

$$u' = Ae^{-5x}.$$

To solve this for u we simply integrate both sides with respect to x

$$u = \int Ae^{-5x}dx.$$

We get a solution for u

$$u = -5Ae^{-5x} + B.$$

We can drop the constants as we have these included in our general solution. This means that we can find our y_2 by substituting back into $y_2 = u(x)e^{2x}$.

$$y_2 = e^{-5x} \cdot e^{2x}$$

giving

$$y_2 = e^{-3x}.$$

Therefore our general solution

$$y = Ae^{2x} + Be^{-3x}.$$

We can use the method of reduction of order to solve linear equations with varying constants such as the Euler equations.

Example: Solve

$$x^2\frac{d^2y}{dx^2} - 6x\frac{dy}{dx} + 6y = 0$$

with solution of $y_1 = x$.

Solution:

We start by making the assumption that $y_2 = xu(x)$. Then

$$y_2' = u'x + u$$

$$y_2'' = u''x + 2u'.$$

Substituting these into our original equation and simplifying we see

$$u''x - 4u' = 0.$$

Letting $w = u'$ and $w' = u''$ we see that we can reduce the order of the differential equation to a first order problem

$$w' = \frac{4}{x}w.$$

This equation is separable so we can take the w terms to the left hand side and the x terms to the right hand side

$$\int \frac{1}{w}dw = \int \frac{4}{x}dx.$$

Solving this for w we get

$$w = Ax^4.$$

As we need an equation for u, not w, we substitute this back in giving

$$u = \frac{A}{5}x^5 + B.$$

Dropping the constants our value for y_2 is

$$y_2 = x^6$$

giving our general solution to be

$$y = Ax + Bx^6.$$

3.4   Uniqueness and Existence of Solutions

When solving a differential equation we do not want to go to all the effort of solving it if our solution is not valid or the equation does not have a solution.

Example: Explain why the differential equation

$$\left(\frac{dy}{dx}\right)^2 + x^2 = -1$$

has no real solutions.

Solution:
What we notice is that the left hand side of the equation is strictly positive due to all the terms being squared. As the left hand side is always positive no real solution can exist as the right hand side is negative.

In this section we will look at a powerful theorem that allows us to find a unique solution over a certain interval for examples that are not as simple as the one we just solved.

3.4.1   Wronskian

Let $x = (x_1, x_2)$ and $y = (y_1, y_2)$ be vectors in the Cartesian plane. If these two vectors are independent then $x \cdot y \neq |x||y|$. That is

$$(x_1^2 + x_2^2)(y_1^2 + y_2^2) \neq (x_1 y_1 + x_2 y_2)$$

Rearranging this gives

$$(x_1 y_2 - x_2 y_1)^2 \neq 0$$

which is the same to $\begin{vmatrix} x_1 & x_2 \\ y_1 & y_2 \end{vmatrix} \neq 0.$

Theorem 5: If $f(x)$ and $g(x)$ are solutions of the homogeneous second order ODE and if $x = \theta$, the vectors $(f(\theta), f'(\theta))$ and $(g(\theta), g'(\theta))$ are linearly independent then every solution can be written as a linear combination of f and g which is called a basis.

If we combine the two ideas above we obtain the Wronskian of the functions y_1 and y_2.

$$W(f, g, x) = \begin{vmatrix} f & f' \\ g & g' \end{vmatrix} = fg' - f'g \neq 0$$

W(f,g,x) is known as the Wronskian of the functions f(x) and g(x). The Wronskian can be used when one solution is known to construct another.

3.5 Real World Applications II

As we did in the second chapter we will look at examples of applications of differential equations, in this section, second order.

Example: A spring is held hanging down with a mass of 2kg attached to it and spring constant $k = 8$. The spring starts at rest at $x = 0$ then is released with downward velocity of $16ms^{-1}$. Model this and solve for time t.

Solution:

Substituting these values above we get the equation

$$2\frac{d^2x}{dt^2} = -8x.$$

From the information given in the question we see that it is released from the point $x = 0$. This means that the first initial condition is $x(0) = 0$. We also know that the initial velocity is $16ms^{-1}$. We know that velocity is the first derivative with respect to time giving our second initial condition as $x'(0) = 16$.

As this is a second order ODE with constant coefficients we can use the auxiliary equation with $a = 2$, $b = 0$ and $c = 8$. From this we see that $2m^2 = -8$ giving $m = \pm 2i$. This produces complex roots so the general solution is

$$x = A\cos(2t) + B\sin(2t)$$

which is the harmonic motion we would expect. Taking the derivative ready for use with the initial conditions

$$\frac{dx}{dt} = -2A\sin(2t) + 2B\cos(2t).$$

Using the initial conditions we obtain the equations

$$0 = A$$
$$16 = 2B$$

where $A = 0$ and $B = 8$. Our final solution is thus

$$x = 8\sin(2t).$$

Plotting the solution to this we can understand what is happening with the spring (check figure 3.3 below). As we release the spring it accelerates downwards under the affect of gravity (remember we took motion down to be positive). This slows as the spring extends eventually coming to a brief stop. This is seen on the graph at the maximum point. Here at the maximum the velocity equals zero. The force exerted by the spring starts to pull the mass in the opposite direction accelerating up, passing the position it was released. Once again as the spring extends too far in this direction it acts against the force slowing it to a halt. As we did not take into consideration energy loss this motion continues forever. If we had taken into account energy loss then the turning points (local maximums and minimums) as time went on would decrease in height until they eventually hit zero.

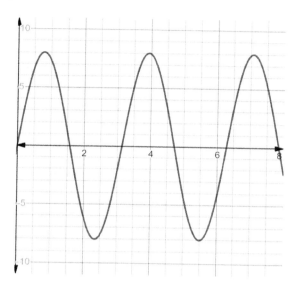

Figure 3.3: The graph of $x = 8sin(2t)$.

3.6 Exercises

1. Find the general solution to

$$\frac{d^2y}{dx^2} + 4\frac{dy}{dx} - 5y = 0.$$

2. Find the general solution to

$$4\frac{d^2y}{dx^2} + 20y = 0.$$

3. Solve

$$\frac{d^2y}{dx^2} + 3\frac{dy}{dx} + 2y = 0$$

with y(0)=1, y'(0)=3.

4. Solve

$$\frac{d^2y}{dx^2} + 5\frac{dy}{dx} = 0$$

with y(0)=2, y'(0)=1.

5. Solve

$$2\frac{d^2y}{dx^2} + 2\frac{dy}{dx} + 5y = sin(x)$$

with y(0)=y'(0)=0.

6. Solve

$$\frac{d^2y}{dx^2} + 4y = x^3$$

with y(0)=2, y'(0)=0.

7. Solve
$$x^2\frac{d^2y}{dx^2} + 6x\frac{dy}{dx} + 6y = 0$$
with y(1)=4, y'(1)=6.

8. Find the general solution to
$$\frac{d^2y}{dx^2} + 4\frac{dy}{dx} + 13y = 0$$

9. Solve
$$\frac{d^2y}{dx^2} + 4\frac{dy}{dx} + 4y = 0$$
with y(0)=1, y'(0)=2

10. Find the general solution to
$$\frac{d^2y}{dx^2} + 9y = 1 + 2x$$

11. Solve
$$x\frac{d^2y}{dx^2} + \frac{dy}{dx} = 0$$
with y(1)=4, y'(1)=3.

Chapter 4

Laplace Transform for ODEs

After the discovery of calculus and the differential equation many mathematicians spent time researching in this area. In the 18th century Euler spent some time looking at integrals of the following form

$$y = \int X(x)e^{ax}dx \tag{4.1}$$

and

$$y = \int X(x)x^{A}dx \tag{4.2}$$

as solutions to differential equations but turned to focus on other areas. Lagrange also used similar integrals within his work on probability density functions. Laplace saw these ideas but instead of looking for a solution that was an integral, used the integral as a transform to take an equation and put it into another form. This transform meant that he could take a differential equation and transform it into a form where he could solve it algebraically.

Laplace transforms are used widely in engineering where they transform functions of time into frequency. We are going to use this convention and look at differentials with respect to time.

4.1   The Laplace Transform

4.1.1   The Laplace Transform

Definition: The Laplace transform $F(s)$ of $f(t)$ for $t > 0$ is

$$F(s) = \mathcal{L}[f(t)] = \int_0^\infty f(t)e^{-st}dt \qquad (4.3)$$

Definition: The inverse Laplace transform $f(t)$ of $F(s)$ for $t > 0$ is

$$f(t) = \mathcal{L}^{-1}[F(s)] = \int_{-\infty}^\infty f(t)e^{-st}dt \qquad (4.4)$$

The definitions above are for a Laplace transform and the inverse transform. We can use equation 4.3 and take a function and transform it. We can then take our transformed equation and use the inverse 4.4 to get back our original function. However what we need to note for these definitions is that it doesn't state when Laplace transform can be used.

Definition: We say that a function $f(t)$ is exponentially bounded if there exists a, n, and K such that

$$|f(t)| \leq Ke^{at} \qquad (4.5)$$

for all $t \geq n$.

Definition: A function is said to be piecewise continuous if it is continuous for all but a finite number of points.

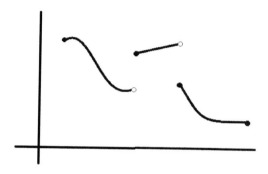

Figure 4.1: Example of a pointwise continuous function.

Theorem 6: Let $f(t)$ be piecewise continuous on $[0, \infty]$ and exponentially bounded. Then the Laplace transform exists such that $\mathcal{L}(t) = F(s)$.

We are now going to look at transforming some key functions that are used reguarly.

Example: Find the Laplace transform $\mathcal{L}[1]$.

Solution:

$$\mathcal{L}[1] = \int_0^\infty e^{-st} dt$$

This integral is an improper integral as the limit is infinity. To rectify this issue we will change the upper limit of the integral to an arbitrary value and take the limit of this value as it tends to infinity.

$$\mathcal{L}[1] = \lim_{n \to \infty} \int_0^n e^{-st} dt$$

We first evaluate the integral

$$\mathcal{L}[1] = \lim_{n \to \infty} \left[-\frac{1}{s} e^{-st} \right]_0^n$$

giving

$$\mathcal{L}[1] = \lim_{n \to \infty} \left[-\frac{1}{s} e^{-sn} \right] + \left[\frac{1}{s} \right].$$

When we take the limit as $n \to \infty$ we see that if $s = 0$ then the transform becomes undefined, when $s < 0$ then the transform is 0, but when $s > 0$ we get a value for the tranform which is

$$\mathcal{L}[1] = \frac{1}{s}.$$

Example: Find the Laplace transform $\mathcal{L}[e^{at}]$.

Solution:

Substituting e^{at} into the transform and replacing the limit we get

$$\mathcal{L}[e^{at}] = \lim_{n \to \infty} \int_0^n e^{at} \cdot e^{-st} dt.$$

As both the bases of the terms in the integral are the same, we can use laws of indices to simplify this.

$$\mathcal{L}[e^{at}] = \lim_{n \to \infty} \int_0^n e^{t(a-s)} dt$$

Taking the integral

$$\mathcal{L}[e^{at}] = \lim_{n \to \infty} \left[\frac{1}{a-s}e^{t(a-s)}\right]_0^n$$

and its limits

$$\mathcal{L}[e^{at}] = \lim_{n \to \infty} \left[\frac{1}{a-s}e^{n(a-s)}\right] - \left[\frac{1}{a-s}\right].$$

Again we need to check where the transform is defined. The only time we where the transform is defined is when $s > a$. This gives

$$\mathcal{L}[e^{at}] = \frac{1}{s-a}$$

Theorem 7: If f(t) and g(t) are functions and a, b scalars then

$$\mathcal{L}[af(t) + bg(t)] = a\mathcal{L}[f(t)] + b\mathcal{L}[g(t)].$$

Proof:
We define the Laplace transform as we did above.

$$\mathcal{L}[af(t) + bg(t)] = \int_0^\infty [af(t) + bg(t)]e^{-st}d$$

Expanding the brackets

$$\mathcal{L}[af(t) + bg(t)] = \int_0^\infty af(t)e^{-st} + bg(t)e^{-st}dt.$$

By the properties of the integral we can split our integral into two.

$$\mathcal{L}[af(t) + bg(t)] = a\int_0^\infty f(t)e^{-st}dt + b\int_0^\infty g(t)e^{-st}dt$$

Which gives

$$\mathcal{L}[af(t) + bg(t)] = a\mathcal{L}[f(t)] + b\mathcal{L}[g(t)].$$

As the use of the Laplace tranform is taking a differentiation problem and turning into an algebraic problem we still need to see how a derivative is transformed.

Example: Find the transform $\mathcal{L}[\frac{dy}{dt}]$.

Solution:

By the definition of the Laplace transform we see that our transform is the integral

$$\mathcal{L}\left[\frac{dy}{dt}\right] = \int_0^\infty \frac{dy}{dt} \cdot e^{-st} dt.$$

As we have an improper integral we are going to change it using the trick we used before

$$\mathcal{L}\left[\frac{dy}{dt}\right] = \lim_{n\to\infty} \int_0^n \frac{dy}{dt} \cdot e^{-st} dt.$$

We are unable to compute the integral directly due to the derivative inside. What we notice is that when we use integration by parts we can 'get rid' of this. As said above, integration by parts is a very useful tool and the reader should keep this at the forefront of their mind.

$$\mathcal{L}\left[\frac{dy}{dt}\right] = \lim_{n\to\infty} [y(t)e^{-st}]_0^n + s \lim_{n\to\infty} \int_0^n y(t) \cdot e^{-st} dt.$$

We are going to assume in this text without proof that $\lim_{n\to\infty} f(n) = 0$. Therefore

$$\mathcal{L}\left[\frac{dy}{dt}\right] = -y(0) + s \lim_{n\to\infty} \int_0^n y(t) \cdot e^{-st} dt.$$

However we know that the second term of the equation above is the Laplace transform of $y(t)$ multiplied by s. This gives the transformation as

$$\mathcal{L}\left[\frac{dy}{dt}\right] = -y(0) + sY(s).$$

where Y(s) is the Laplace transform of $y(t)$.

Using the above transformation we can find the transform of the second derivative and the nth derivative. If we repeat the process above with the second derivative we see that

$$\mathcal{L}\left[\frac{d^2 y}{dt^2}\right] = -y'(0) + s \lim_{n \to \infty} \int_0^n \frac{dy}{dt} \cdot e^{-st} dt.$$

Completing the integration we find the Laplace transform of the second derivative.

$$\mathcal{L}\left[\frac{d^2 y}{dt^2}\right] = -y'(0) - sy(0) + s^2 F(s).$$

Noticing the pattern above gives the transform of the nth derivative.

$$\mathcal{L}\left[y^{[n]}(x)\right] = s^n F(s) - s^{n-1} y(0) - s^{n-2} y'(0) - \ldots - y^{[n-1]}(0).$$

Here we have used the square brackets in superscript to note a derivative of higher order.

4.1.2 Use of Tables

One of the powers of using Laplace transforms is that it makes difficult calculations easier to deal with. That said it would be tiresome to compute the Laplace transform of each term every time we want to solve an equation. Laplace transforms are known for a lot of function and alongside the theorem above we can quickly and easily transform functions.

f(t)	F(s)
1	$\frac{1}{s}$
t^n	$\frac{n!}{s^{n+1}}$
$\sqrt{t}$	$\frac{\sqrt{\pi}}{2s^{3/2}}$
$cos(at)$	$\frac{s}{s^2+a^2}$
$sin(at)$	$\frac{a}{s^2+a^2}$
$tcos(at)$	$\frac{s^2+a^2}{(s^2+a^2)^2}$
$tsin(at)$	$\frac{2as}{(s^2+a^2)^2}$
e^{at}	$\frac{1}{s-a}$
$e^{at}j(t)$	$J(s-a)$
$t^n e^{at}$	$\frac{n!}{(s-a)^{n-1}}$
$y(t)$	$Y(s)$
$y'(t)$	$sY(s) - y(0)$
$y^{[n]}(t)$	$s^n Y(s) - s^{n-1}y(0) - ... - y^{[n-1]}(0)$

We need to be able to use tables to transform equations and expressions.

Example: Find the Laplace transform of $cos(3t) + t^3 + 2e^{5t}$.

Solution:

Using the table we see that

$$\mathcal{L}[cos(3t)] = \frac{s}{s^2 + 9},$$

$$\mathcal{L}[t^3] = \frac{3!}{s^4},$$

$$\mathcal{L}[2e^{5t}] = \frac{1}{s - 5}.$$

Putting these together we obtain for the above expression

$$\left[\frac{s}{s^2+9}\right] + \left[\frac{3!}{s^4}\right] + 2\left[\frac{1}{s-5}\right]$$
$$= \frac{s}{s^2+9} + \frac{6}{s^4} + \frac{2}{s-5}.$$

Example: Find the inverse transform of $\frac{4}{s} - \frac{2}{s^2+4} + g(s-5)$

Solution:

There is three terms and we can convert each one back individually.

$$\mathcal{L}^{-1}\left[\frac{4}{s}\right] = 4$$
$$\mathcal{L}^{-1}\left[-\frac{2}{s^2+4}\right] = -sin(2t)$$
$$\mathcal{L}^{-1}\left[g(s-5)\right] = e^{5t}g(t)$$

If we use the table and convert the second term we get $-sin(2t)$. However if we bring the negative to the top of the fraction (and the square of -2 is 4) we convert this back to $sin(-2t)$. As sine is an odd function this makes sense as the two terms are equal. Putting these together we see that our complete inverse transformation is

$$4 - sin(2t) + e^{5t}g(t)$$

4.2 Heaviside and Dirac Delta Functions

Before moving on to solving differential equations we are going to look at two functions that often appear in Laplace transform problems. These are the Heaviside and Dirac delta functions. Both of these are examples of step functions.

Definition: (Step Function) A function on the reals $\mathbb{R}$ that can be written as a finite combinations of semi-open intervals is a step function.

4.2.1   Heaviside Function

The Heaviside function is named after Oliver Heaviside, born in Middlesex England. Heaviside man great strides in using complex numbers in electrical circuits. This can be seen in the function named after him.

The Heaviside function is a simple piecewise function over $\mathbb{R}$. The function represents a signal that switches on at a certain point and then stays on. We can see this when we write the function mathematically.

$$H(x - c) = \begin{cases} 0 & x \leq c \\ 1 & x > c \end{cases} \tag{4.6}$$

where c is the point where the switch turns on.

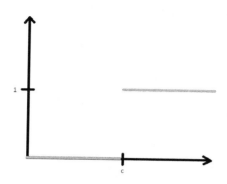

Figure 4.2: An example of a Heaviside function.

We need to make a clear difference between two different notation:

$$H(x - 2) \neq 2H(x). \tag{4.7}$$

The first is the function that is zero up to $x = 2$ where it then takes a value of 1. The second function is zero up to $x = 0$ where it then takes a value of 2. See below for a side by side comparison of the two functions in 4.7.

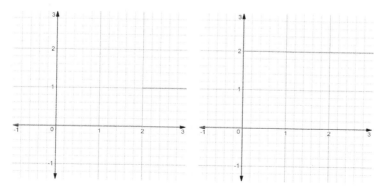

4.2.2   Dirac Delta Function

The Dirac Delta function is similar the Heaviside function in many ways. Named after English physicist Paul Dirac the function is indefinitely zero apart from a single point of infinite mass.

$$\delta(x - c) = \begin{cases} +\infty & x = c \\ 0 & x \neq c \end{cases} \tag{4.8}$$

which also satisfies

$$\int_{-\infty}^{\infty} \delta(x - c)dx = 1. \tag{4.9}$$

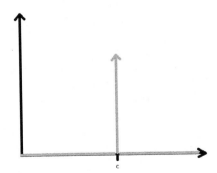

Figure 4.3: An example of a Dirac Delta function.

What is also of note is that the derivative with respect to x of $H(x - c)$ is the $\delta(x - c)$. The proof is this is omitted from this book as it uses distributional derivative.

$$H(x - c) = \int_{-\infty}^{\infty} \delta(x - c)dx \tag{4.10}$$

We can now add the Laplace transforms of the Heaviside and Dirac Delta function into our table.

f(t)	F(s)
$H(x-c)$	$\frac{e^{cs}}{s}$
$\delta(x-c)$	e^{cs}

4.3   Solving Transformed ODEs

The power of the Laplace transform is to convert linear ODEs into an algebraic problems. This makes the problem far easier to solve and use of tables speeds up the process. One key algebraic skill that will be used in most problems is being able to deal with and manipulate partial fractions.

4.3.1   Partial Fractions

When adding fractions we take two terms and make them into a single, often more complicated, term. When dealing with Laplace transforms we often need to convert back fractions into multiple simpler ones. Sometimes these fractions are complicated but we can use a method called partial fraction decomposition. This separates fractions into two or more simpler fractions, which means we can use Laplace tables directly to tranform these back directly.

Example: Find the partial fraction decomposition of

$$\frac{6x+1}{x^2+3x+2}$$

Solution:

Our first step is to factorise the denominator of the fraction.

$$\frac{6x + 1}{(x + 2)(x + 1)}$$

We can express the fraction above as two fractions with $(x + 2)$ as the denominator of one of the fractions and $(x + 1)$ as the other. At this moment in time we don't know the numerator of either. To work out the numerator we are going to equate our new fractions with the original one.

$$\frac{6x + 1}{(x + 2)(x + 1)} = \frac{A}{(x + 2)} + \frac{B}{(x + 1)}$$

Multiplying through by $(x + 2)(x + 1)$ we get

$$6x + 1 = A(x + 1) + B(x + 2)$$

The trick to find the unknowns A and B is to substitute in -1 and -2. The reason for this is that each of these will cause one of A and B to disappear meaning we can solve for the other.

Let $x = -1$

$$6(-1) + 1 = B(1)$$

giving $B = -5$.

Let $x = -2$

$$6(-2) + 1 = A(-1)$$

giving $A = 11$.

With these values for A and B we obtain our decomposition as

$$\frac{11}{(x + 2)} - \frac{5}{(x + 1)}.$$

There is a few key points we need to be careful of.

1. If we have a value to any power other than 1 we need to include all lower powers.

$$\frac{2x+1}{(x+1)^3} = \frac{A}{(x+1)} + \frac{B}{(x+1)^2} + \frac{C}{(x+1)^3}$$

2. If we can factorise the denominator we should. Note that $x^2 - 4 = (x+2)(x-2)$ (ie a difference of two squares) but we cannot factorise $x^2 + 4$.

3. If you have a quadratic factor (that won't factorise) then the partial fraction we are looking for is

$$\frac{Ax+B}{ax^2+bx+c}$$

4.3.2 Solving ODEs Using Laplace Transform

We can now finally get down to business and solve ODEs using Laplace transforms. Of particular importance when solving is the use of the tables and partial fractions.

Example: Solve the equation below by using the method of Laplace transform.

$$\frac{d^2y}{dt^2} - 4y = e^{3t}$$

with initial conditions $y(0) = y'(0) = 0$.

Solution:

We have already seen methods to solve ODEs in this form so you can check that our answer agrees using other methods.

Remember that we can take the transform of each term individually. So for the Laplace transform we get

$$s^2Y(s) - sy(0) - y'(0) - 4Y(s) = \frac{1}{s-3}.$$

When transforming the derivatives we end up with terms using the initial conditions. To be able to solve the problem we need to substitute these. We are lucky with this example the initial conditions says that these terms are equal to zero. This makes the calculations for this ODE simpler.

$$s^2 Y(s) - 4Y(s) = \frac{1}{s-3}$$

Now the initial conditions have been substituted we can solve for $Y(s)$. We start by factorising out $Y(s)$

$$Y(s)(s^2 - 4) = \frac{1}{s-3}.$$

Dividing both sides by $(s^2 - 4)$ we obtain an expression for $Y(s)$

$$Y(s) = \frac{1}{(s^2 - 4)(s - 3)}$$

As $(s^2 - 4)$ is a difference of two squares we can factorise this.

$$Y(s) = \frac{1}{(s-2)(s+2)(s-3)}$$

There is no transform in our tables for a fraction like the one we have on the right hand side. This is when we have to use partial fraction decomposition on the right had side. This will hopefully get this into a form where we can use the tables directly.

$$\frac{1}{(s-2)(s+2)(s-3)} = \frac{A}{s-2} + \frac{B}{s+2} + \frac{C}{s-3}$$

Multiplying through by $(s-2)(s+2)(s-3)$

$$1 = A(s+2)(s-3) + B(s-2)(s-3) + C(s-2)(s+2).$$

By letting $s = 3$ we see that $C = \frac{1}{5}$, letting $s = 2$ we see that $A = -\frac{1}{4}$ and, letting $s = -2$ we finally see that $B = \frac{1}{20}$. This give

$$Y(s) = \frac{1}{5} \cdot \frac{1}{s-3} - \frac{1}{4} \cdot \frac{1}{s-2} + \frac{1}{20} \cdot \frac{1}{s+2}.$$

Using the table we can convert back for our final solution

$$y(t) = \frac{1}{5}e^{3t} - \frac{1}{4}e^{2t} + \frac{1}{20}e^{-2t}.$$

If you want to check this, use the content from chapter 3 and see if the answers match.

Example: Solve the equation below using Laplace transform.

$$\frac{d^2y}{dx^2} - \frac{dy}{dx} = sin(4t)$$

with initial conditions $y(0) = 0$ and $y'(0) = 1$.

Solution:

We start by taking the Laplace transform and immediately substituting in the initial conditions

$$s^2Y(s) - sY(s) - 1 = \frac{4}{s^2 + 16}.$$

Solving for Y(s)

$$Y(s) = \frac{4}{(s^2 + 16)(s^2 - s)} + \frac{1}{s^2 - s}$$

To be able to use the method of partial fraction decomposition we need to have only one fraction on the right hand side. Simply adding the two fractions solves this problem.

$$Y(s) = \frac{s^2 + 20}{(s^2 + 16)(s^2 - s)}$$

Now we will use partial fraction decomposition noting that we will use key point number three from the previous section.

$$\frac{s^2 + 20}{(s^2 + 16)(s)(s - 1)} = \frac{A}{s} + \frac{Bs + C}{s^2 + 16} + \frac{D}{s - 1}$$

Multiplying through by the denominator of the left hand side

$$s^2 + 20 = A(s-1)(s^2+16) + (Bs+C)(s)(s-1) + D(s)(s^2+16)$$

By letting $s = 1$ we get $D = \frac{21}{17}$ and letting $s = 0$ we get $A = -\frac{5}{4}$. As the second term has two unknowns we will have to substitute two values for s and solve the pair of simultaneous equations. By substituting $s = -1$ and $s = 2$ we get the pair of equations which we can then solve for B and C. We see that $B = 1/68$ and $C = -16/68$. Putting these back into our equation

$$Y(s) = -\frac{5}{4} \cdot \frac{1}{s} + \frac{1}{68} \cdot \frac{s}{s^2+16} - \frac{4}{68} \cdot \frac{4}{s^2+16} + \frac{21}{17} \cdot \frac{1}{s-1}$$

Using the table to convert back to our final answer

$$y(t) = -\frac{5}{4} + \frac{1}{68}cos(4t) - \frac{1}{17}sin(4t) + \frac{21}{17}e^t$$

4.4   Nonconstant Coefficients

In the previous sections of this chapter we have only seen differential equations where there are constant coefficients. Now we are going to look at variable coefficients. For this we are going to add one more transform to our table.

f(t)	F(s)
$t^n y(t)$	$(-1)^n Y^{[n]}(s)$

Now we have that $t^n y(t)$ transforms into $(-1)^n Y^{[n]}(s)$ we can look at examples.

Example: Solve the equation below using Laplace transform.

$$t\frac{d^2y}{dt^2} + t\frac{dy}{dt} - y = 2$$

with initial conditions $y(0) = -2$ and $y'(0) = 4$.

Solution:

Taking the transform of the first term we see

$$\mathcal{L}\left[t\frac{d^2y}{dt^2}\right] = -\frac{d}{ds}\mathcal{L}\left[\frac{d^2y}{dt^2}\right]$$
$$= -\frac{d}{ds}[s^2Y(s) - sy(0)]$$
$$= -s^2Y'(s) - 2sY(s) - y(0)$$

and for the second

$$\mathcal{L}\left[t\frac{dy}{dt}\right] = -\frac{d}{ds}\mathcal{L}\left[\frac{dy}{dt}\right]$$
$$= -\frac{d}{ds}\mathcal{L}[sY(s) - y(0)]$$
$$= -sY'(s) - Y(s).$$

Using these we transform the whole equation and with $y(0) = -2$ to get

$$s^2Y'(s) + 2sY(s) + sY'(s) + 2Y(s) = -\frac{2 + 2s}{s}.$$

This is a first order differential equation. We need to use the methods in the second chapter to solve this. Looking at the form this is in we are going to solve this by using an integrating factor.

So far we have only used one initial condition. This is because when we differentiated we lost one of the conditions. We will use the second condition later.

We start by factorising the coefficients of $Y'(s)$ and $Y(s)$

$$Y'(s)[s^2 + s] + Y(s)[2s + 2] = -\frac{2 + 2s}{s}.$$

Dividing through by $[s^2 + s]$ we will get the equation into the form

$$\frac{dy}{dx} + P(x)y = Q(x).$$

Here our $P(x) = \frac{2}{s}$ and $Q(x) = -\frac{2}{s^2}$. Thus we obtain

$$Y'(s) + \frac{2}{s}Y(s) = -\frac{2}{s^2}.$$

Now we use the integrating factor

$$\mu = e^{\int 2/s}$$

to get

$$s^2 Y(s) = -\int \frac{2}{s^2} s^2 ds.$$

Therefore

$$Y(s) = -\frac{2}{s} + \frac{c}{s^2}.$$

Fortunately this example does not need to use partial fractions. So transforming back gives

$$y(t) = -2 + ct.$$

Finally we use our second initial condition of $y'(0) = 4$ and see that $c = t$. Our final solution is

$$y(t) = -2 + 4t.$$

4.5   Exercises

1. Use partial fraction decomposition on

$$\frac{3x + 4}{x(x^2 - 5x + 6)}.$$

2. Show that $\mathcal{L}[sin(at)] = \frac{a}{s^2+a^2}$,

3. Use the Laplace tables to transform

$$4e^{5t} - cos(2t) + t^4e^{3t}.$$

4. Solve

$$\frac{d^2y}{dx^2} - y = cos(x)$$

using a Laplace transform with y(0)=y'(0)=0.

5. Show directly that the Laplace transform of e^t is $\frac{1}{s-1}$.

6. Solve

$$\frac{dy}{dt} + y = t$$

using a Laplace transform with y(0)=1.

7. Solve

$$\frac{d^2y}{dt^2} - 3\frac{dy}{dt} + 2y = e^{3t}$$

with y(0)=1, y'(0)=0.

8. Solve
$$\frac{dy}{dt} - 2y = e^{3t}$$
with y(0)=-5.

9. Solve
$$\frac{dy}{dx} = xe^x + 2x^x + t$$
with y(0)=3.

Chapter 5

Higher Order ODEs

So far we have looked only at first and second order ODEs. These appear regularly in the real world and we have looked at certain examples of these. Now we are going to move on to an nth order linear homogeneous equation.

$$a_n y^{[n]} + a_{n-1} y^{[n-1]} + ... a_0 = 0 \tag{5.1}$$

where $y^{[n]}$ is the nth derivative. With the aid of the principle of superposition the general solution is

$$y = a_1 y_1 + a_2 y_2 + ... a_n y_n. \tag{5.2}$$

These are n linearly independent solutions of the equation. The linearly independence of the solutions can be found by their Wronskian.

5.1 Guassian Elimination

When solving differential equations we may need to solve systems of linear equations. If these systems contain multiple unknowns solving them becomes tricky. If we were to try and solve this using substitution it would be a long and tedious affair. A method named after Guass (although it was used far earlier by Asian mathematicians) uses matrices to find the unknowns which greatly reduces the calculations needed. This method will be key later on in this chapter.

Let x, y, and z be three unknowns and a to l be real values then

$$ax + by + cz = j$$
$$dx + ey + fz = k$$
$$gx + hy + iz = l$$

becomes $\begin{vmatrix} a & b & c & j \\ d & e & f & k \\ g & h & i & l \end{vmatrix}$.

The first column represents the coefficients of the x values, the second column is the coefficients of y, and the third is the coefficient of z.

At its heart, Guassian elimination is an algorithm. The algorith is as follows:

1. Write the system of equations as a matrix.

2. Divide the first row by the coefficient of the first row first column so that this value is one.

3. Subtract multiples of the first row away from the following rows such that it cancels their terms in the first rows.

4. Repeat the process above for the sequential rows such that we have an upper triangular matrix.

5. Working backwards, subtract multiples of the rows to obtain a diagonal matrix. This will subsequently give values for each of the unknowns.

One thing to note is that we can switch the equations around to make the calculations easier.

Example: Solve the system of equations

$$2x + 3y + z = 4.5$$
$$5x + y + 2z = 8.5$$
$$3x + 3y + 3z = 4.5$$

Solution:

First step; we put our linear equations into a matrix. We take the first column to be our coefficients of x, second y and third z.

$$\begin{bmatrix} 2 & 3 & 1 & | & 4.5 \\ 5 & 1 & 2 & | & 8.5 \\ 3 & 3 & 3 & | & 4.5 \end{bmatrix}$$

Step 2; we are going to divide the first row by 2 such that we get 1 in the top left corner.

$$\begin{bmatrix} 1 & 1.5 & 0.5 & | & 2.25 \\ 5 & 1 & 2 & | & 8.5 \\ 3 & 3 & 3 & | & 4.5 \end{bmatrix}$$

Step 3; subtracting 5 lots of the first row away from the second and 3 lots of the first row away from the third gives

$$\begin{bmatrix} 1 & 1.5 & 0.5 & | & 2.25 \\ 0 & -6.5 & -0.5 & | & -2.75 \\ 0 & -1.5 & 1.5 & | & -2.25 \end{bmatrix}.$$

Step 4; repeat the process for all our rows, so we divide our second row by -6.5.

$$\begin{bmatrix} 1 & 1.5 & 0.5 & 2.25 \\ 0 & 1 & 1/13 & 11/26 \\ 0 & -1.5 & 1.5 & -2.25 \end{bmatrix}$$

Adding 1.5 lots of row 2 to row 3

$$\begin{bmatrix} 1 & 1.5 & 0.5 & 2.25 \\ 0 & 1 & 1/13 & 11/26 \\ 0 & 0 & 21/13 & -21/13 \end{bmatrix}.$$

Now we have an upper triangular matrix we can start to form a diagonal matrix. Subtracting 1.5 lots of row 2 away from row 1.

$$\begin{bmatrix} 1 & 0 & 5/13 & 21/13 \\ 0 & 1 & 1/13 & 11/26 \\ 0 & 0 & 21/13 & -21/13 \end{bmatrix}$$

Dividing row 3 by $\frac{21}{13}$ we get

$$\begin{bmatrix} 1 & 0 & 5/13 & 21/13 \\ 0 & 1 & 1/13 & 11/26 \\ 0 & 0 & 1 & -1 \end{bmatrix}.$$

Finally we will subtract $\frac{1}{13}$ lots of row 3 away from row 2 and $\frac{5}{13}$ lots of row 3 away from row 1.

$$\begin{bmatrix} 1 & 0 & 0 & 2 \\ 0 & 1 & 0 & 0.5 \\ 0 & 0 & 1 & -1 \end{bmatrix}$$

This gives the answers to the system of the linear equations to be $x = 2$, $y = 0.5$ and $z = -1$.

5.2　Auxiliary Equation

To solve higher order linear homogeneous ODEs we can use the auxiliary as we did with second order.

$$a_n m^n + a_{n-1} m^{n-1} + ... a_0 = 0 \tag{5.3}$$

As with second order distinct real, complex and repeated roots give solution as they did before. There is one difference we have to be careful of. As an equation with order n, we have n independent solutions. These may have repeated, real or complex solutions. If we have n repeated real solutions then $y = e^{mx}$, $y = xe^{mx}$, $y = x^2 e^{mx}$, ..., $y = x^n e^{mx}$. If we have n repeated complex solutions with $m = a \pm bi$ then $y = e^{ax} \cos(bx) + e^{ax} \sin(bx)$, $y = xe^{ax} \cos(bx) + xe^{ax} \sin(bx)$, $y = x^2 e^{ax} \cos(bx) + x^2 e^{ax} \sin(bx)$, ..., $y = x^{n-1} e^{ax} \cos(bx) + x^{n-1} e^{ax} \sin(bx)$.

Let us look at some examples. We are going to use Lagrange's notation for the following examples.

Example: Solve

$$y^{[3]} + y' = 0$$

with initial conditions of $y(0) = 0$, $y'(0) = 1$ and $y''(0) = 3$.

Solution:

Using the auxiliary equation we obtain $m(m^2 + 1) = 0$ giving $m = 0$ and $m = \pm i$. This gives the general solution as

$$y = A + B\cos(x) + C\sin(x)$$

Using the initial conditions we see

$$0 = A + B$$
$$1 = C$$
$$3 = -B$$

With a system of three simultaneous equations we are going to put these into a matrices and use row reduction (also known as Guassian elimination) to solve for the constants. This may not be necessary for this example but we show it as for extra practise for the reader.

$$\begin{bmatrix} 1 & 1 & 0 & | & 0 \\ 0 & 0 & 1 & | & 1 \\ 0 & -1 & 0 & | & 3 \end{bmatrix}$$

For row reduction we need a single entry of 1 in each of the coulombs and a single entry of 1 in each of the rows. To make this easier on ourselves we are going to switch row 2 and 3 around as well as multiplying row 2 by -1.

$$\begin{bmatrix} 1 & 1 & 0 & | & 0 \\ 0 & 1 & 0 & | & -3 \\ 0 & 0 & 1 & | & 1 \end{bmatrix}$$

We now subtract row 2 away from row 1

$$\begin{bmatrix} 1 & 0 & 0 & | & 3 \\ 0 & 1 & 0 & | & -3 \\ 0 & 0 & 1 & | & 1 \end{bmatrix}$$

This gives values for our constants; $A = 3$, $B = -3$, and $C = 1$. Putting these into our general solution gives

$$y = 3 - 3cos(x) + sin(x)$$

In the last example we had $m(m + 1)^2 = 0$. This gave us three different roots; $m = 0$, $m = \pm i$. Like with second order equations, higher order equations we may have repeated roots but repeated complex roots. Like with repeated roots for second order, each time we multiply by x.

5.3   Laplace Transform for Higher Order ODEs

Another method to solve higher order linear ODEs is to use Laplace transforms. We saw that in chapter 4 that we can transform any order derivative. We will use this in the further examples.

Example: Solve the fourth order ODE using Laplace transform.
$$y^{[4]} - 2y'' + y = cos(2t)$$
with initial conditions $y(0) = y'(0) = y''(0) = y'''(0) = 0$.

Solution:

Taking the Laplace transform and using the initial conditions we obtain
$$s^4 Y(s) - 2Y(s) + Y(s) = \frac{s}{s^2 + 4}.$$
Factoring out Y(s) and dividing through by the bracket gives
$$Y(s) = \frac{s}{(s^4 - 1)(s^2 + 4)}.$$
The first bracket is a difference of two squares so we can expand this giving us a form we can do partial fraction decomposition with.
$$Y(s) = \frac{s}{(s - 1)^2(s + 1)^2(s^2 + 4)}$$
Now we start the decomposition. We need to remember the key points we saw in the previous chapter.
$$\frac{s}{(s - 1)^2(s + 1)^2(s^2 + 4)} = \frac{A}{s + 1} + \frac{B}{(s + 1)^2} + \frac{C}{s - 1} + \frac{D}{(s - 1)^2} + \frac{Es + F}{s^2 + 4}$$
This partial fraction problem requires Guassian elimination that we saw in the previous section. For conciseness we will skip the algebra and just quote the values of the constants. $A = -\frac{1}{50}$, $B = -\frac{1}{20}$, $C = -\frac{1}{50}$, $E = \frac{1}{20}$ and, F=0. This gives
$$Y(s) = -\frac{1}{50} \cdot \frac{1}{s + 1} - \frac{1}{20} \cdot \frac{1}{(s + 1)^2} - \frac{1}{50} \cdot \frac{1}{s - 1} + \frac{1}{20} \cdot \frac{1}{(s - 1)^2} + \frac{1}{25} \cdot \frac{s}{s^2 + 4}.$$
Transformer back gives our full solution
$$y(t) = -\frac{1}{50}e^{-t} - \frac{1}{20}te^{-t} - \frac{1}{50}e^{t} + \frac{1}{20}te^{t} + \frac{1}{25}cos(2t).$$

5.4   Reduction of Order for Higher Order

We can also use reduction of order for higher order equations to help us solve them. This may be slightly less useful for order of four or above as the problem may still be complicated but it will still make the problem easier to solve.

Example: Find the general solution to

$$y^{[3]} - 8y = 0$$

with the solution $y_1 = e^{2x}$.

Solution:

Assuming $y_2 = u(x)e^{2x}$ we find the derivatives

$$y_2' = u'e^{2x} + 2ue^{2x},$$
$$y_2'' = u''e^{2x} + 4u'e^{2x} + 4ue^{2x},$$
$$y_2^{[3]} = u^{[3]}e^{2x} + 6u''e^{2x} + 12u'e^{2x} + 8ue^{2x}.$$

Substituting these into our original equation and simplifying we obtain

$$u^{[3]}e^{2x} + 6u''e^{2x} + 12u'e^{2x} = 0.$$

Letting $w = u'$ and $w' = u''$ and simplifying

$$w'' + 6w' + 12w = 0.$$

Using the auxiliary equation we obtain the general solution for w of

$$w = Ae^{-3x}sin(\sqrt{3}x) + Be^{-3x}cos(\sqrt{3}x).$$

We substitute back u and integrate (we have not given values to the constants as we will drop them soon)

$$u = C[e^{-3x}sin(\sqrt{3}x)] + D[e^{-3x}cos(\sqrt{3}x)].$$

giving

$$y_2 = e^{-x}sin(\sqrt{3}x) + e^{-x}cos(\sqrt{3}x).$$

Our general solution therefore is

$$y = Fe^{2x} + Ge^{-x}sin(\sqrt{3}x) + He^{-x}cos(\sqrt{3}x)$$

5.5 Exercises

1. Use Guassian elimination to solve

$$A + 3C = -3$$
$$B + C = -2$$
$$2A + 2B + 2C = 2$$

2. Find the general solution to

$$y^{[3]} - y'' - y' + y = 0$$

3. Find the general solution to

$$y^{[3]} - 4y'' + 5y' - 2y = 0.$$

4. Solve
$$y^{[3]} + 3y'' + 3y' + y = 0$$
with y(0)=y'(0)=0, y"(0)=5.

5. Solve
$$y^{[4]} - 2y'' + y = 0$$
with y(0)=1, y'(0)=2, y"(0)=3, y"'(0)=4.

6. Find the general solution to
$$y^{[5]} + 5y^{[4]} - 2y^{[3]} - 10y'' + y' + 5y = 0.$$

7. Find the general solution to
$$y^{[3]} + y' = cos(2x).$$

8. Find the general solution to
$$y^{[3]} + y' = e^x.$$

Chapter 6

Series Solutions

In previous chapters we have seen specific methods for certain differential equation. Now we are going to look at a more general method. The idea behind this idea is that you can write the solution as a power series

$$y = \sum_{n=0}^{\infty} a_n x^n.$$

With this we can differentiate and substitute these for the parts of the differential equations we are looking at.

We are going to look at two differential equations; Airy's equation and the Hermite differential equation. Before we look at these we need to see how to change the index of a summation.

6.1 Changing Index of Summations

Later as we shall see, when we solve differential equations with a power series, we will often get summations which start at different points. This becomes a problem when we want to simplify and go on to get a solution that is usable.

Example: Change the index of the summation so it starts at $n = 2$.

$$\sum_{n=0}^{\infty} a_n x^n$$

Solution:

For the index of this summation to start at $n = 2$ we need to take out the $n = 0$ and $n = 1$ parts of the summation and add these on to the expression. Remember that a summation is a collection of additions. First we take out the $n = 0$ part from the summation. We need to replace all parts in the summation that include n with 0. Doing this we get

$$\sum_{n=0}^{\infty} a_n x^n = a_0 + \sum_{n=1}^{\infty} a_n x^n.$$

We get a single a value as x^0 is 1. We follow this step by taking out the $n = 1$ part of the summation.

$$\sum_{n=0}^{\infty} a_n x^n = a_0 + a_1 x + \sum_{n=2}^{\infty} a_n x^n.$$

In the example above we see that we have changed the starting point of our summation. Here we only have one term that has a summation in it. In the coming examples in order to be able to simplify our work and get to a solution we will need to change the powers of x which means we will have to take a different approach. This will then allow us to take multiple summations and simplify them into a single one.

Example: Change the index of the summation of the following expression and simplify into one summation.

$$\sum_{n=1}^{\infty} a_n x^{n-1} + \sum_{n=0}^{\infty} a_n x^n$$

Solution:

As our primary goal is to take these two summations and simplify into one where we have only one power of x, we need to focus on these and somehow make them the same. We do this by letting $k = n - 1$ in the first summation and $k = n$ in the second. Substituting these into the expression above we get

$$\sum_{n=1}^{\infty} a_n x^{n-1} + \sum_{n=0}^{\infty} a_n x^n = \sum_{k=0}^{\infty} a_{k+1} x^k + \sum_{k=0}^{\infty} a_k x^k.$$

Straight away we see that both summations start and finish at the same point so we can write this as a single summation.

$$\sum_{n=1}^{\infty} a_n x^{n-1} + \sum_{n=0}^{\infty} a_n x^n = \sum_{k=0}^{\infty} a_{k+1} x^k + a_k x^k.$$

Both terms in the summation have x^k as part of them so we can factorise these out and complete the simplification.

$$\sum_{n=1}^{\infty} a_n x^{n-1} + \sum_{n=0}^{\infty} a_n x^n = \sum_{k=0}^{\infty} (a_{k+1} + a_k) x^k$$

As the right hand side of the equation only has k as part of the summation we can rename this n for the sake of consistency.

We were fortunate in this example that after making the substitution we could go straight on to simplify into a single summation.

This may not always be the case.

Example: Change the index of the summation of the following
expression and simplify into one summation.

$$\sum_{n=1}^{\infty} a_n x^n + \sum_{n=2}^{\infty} a_n x^{n-2}$$

Solution:

We start by looking at the powers of x and trying to make these
consistent between the two summations. This makes our choices
$k = n$ in the first summation and $k = n - 2$ in the second. Substituting these in we get

$$\sum_{n=1}^{\infty} a_n x^n + \sum_{n=2}^{\infty} a_n x^{n-2} = \sum_{k=1}^{\infty} a_k x^k + \sum_{k=0}^{\infty} a_{k+2} x^k$$

Now we have the same power of x but we have an additional problem
that the summations start at different points. This is where the first
example we did in this chapter comes in handy. We can pull out the
$k = 0$ part of the second summation.

$$\sum_{n=1}^{\infty} a_n x^n + \sum_{n=2}^{\infty} a_n x^{n-2} = a_2 + \sum_{k=1}^{\infty} a_k x^k + \sum_{k=1}^{\infty} a_{k+2} x^k$$

Now we have the summations starting at the same point we can
proceed as we did in the previous example.

$$\sum_{n=1}^{\infty} a_n x^n + \sum_{n=2}^{\infty} a_n x^{n-2} = a_2 + \sum_{k=1}^{\infty} a_k x^k + a_{k+2} x^k$$

$$= a_2 + \sum_{k=1}^{\infty} (a_k + a_{k+2}) x^k$$

We cannot simplify this any further and have to leave the a_2 as a
seperate term in front of the summation.

6.2 Solving ODE's with Power Series

As we said at the start of this chapter, the essence of solving a differential equation with a power series is assuming the solution can be written as

$$y = \sum_{n=0}^{\infty} a_n x^n.$$

By taking the derivative of this and substituting into our original problem we can solve for the specific y our ODE has.

We are going to look at a couple of examples. The first being a standard example looked at in a lot of mathematical texts, Airy's equation. This equation takes the form of

$$\frac{d^2 y}{dx^2} \pm k^2 x y = 0.$$

Conventional we let $k^2 = 1$ and only take the negative part of the $\pm$.

The second example we will look at is a slightly more abstract example called the Hermite differential equation. This differential equation of the form

$$\frac{d^2 y}{dx^2} - 2x \frac{dy}{dx} + \lambda y = 0$$

is also known to have a singularity at ∞. We shall not look at that detail specifically but it is interesting to note.

Example: (Airy's equation) Solve the differential equation below by power series.

$$\frac{d^2 y}{dx^2} - xy = 0$$

Solution:

As stated above we let y be represented by a power series. We need
to take note that when we take derivatives the power of x decreases
by one. This means that if we do not change where the summation
starts we will get a negative power of x, which we do not want. Thus
for the first derivative of x we start our summation at $n = 1$ and for
the second we start at $n = 2$. Doing this gives us

$$y = \sum_{n=0}^{\infty} a_n x^n,$$

$$\frac{dy}{dx} = \sum_{n=1}^{\infty} n a_n x^{n-1},$$

$$\frac{d^2 y}{dx^2} = \sum_{n=2}^{\infty} n(n-1) a_n x^{n-2}.$$

Taking the y and $\frac{d^2 y}{dx^2}$ we substitute into Airy's equation. We see
that the y term is being multiplied by x so the power of the x inside
the summation will increase by 1.

$$\sum_{n=2}^{\infty} n(n-1) a_n x^{n-2} - \sum_{n=0}^{\infty} a_n x^{n+1} = 0$$

We run into the problem we spoke of earlier, the powers of x inside
the summations do not match. We therefore have to take $k = n - 2$
and $k = n + 1$ for the first and second summation respectively. This
will enable the x terms to now have the same power.

$$\sum_{k=0}^{\infty} (k+2)(k+1) a_{k+2} x^k - \sum_{k=1}^{\infty} a_{k-1} x^k = 0$$

Now we have the same power of x but we do not have our summations starting at the same point. This means that we need to pull out

the $k = 0$ part of the first summation. Doing this and simplifying gives

$$2a_2 + \sum_{k=1}^{\infty}(k+2)(k+1)a_{k+2}x^k - \sum_{k=1}^{\infty}a_{k-1}x^k = 0$$

$$2a_2 + \sum_{k=1}^{\infty}(k+2)(k+1)a_{k+2}x^k - a_{k-1}x^k = 0$$

$$2a_2 + \sum_{k=1}^{\infty}\left((k+2)(k+1)a_{k+2} - a_{k-1}\right)x^k = 0$$

From the equation above we come to two conclusions, either the term outside the summation equals zero, or the expression multiplying the x equals zero. This gives us two equations:

$$2a_2 = 0,$$
$$(k+2)(k+1)a_{k+2} - a_{k-1} = 0.$$

Straight away we can see that $a_2 = 0$ and with some rearranging of the second we come to what is called a recurrence relation (sometimes called a difference equation). This is where an equation is defined by a previous term. Therefore for the other equation we observe

$$a_{k+2} = \frac{a_{k-1}}{(k+2)(k+1)}.$$

With this recurrence relation we found above we can give more information for the power series for y. Letting $k = 0, 1, 2, 3...$ we get

$$a_2 = 0$$

$$a_3 = \frac{a_0}{2 \times 3}$$

$$a_4 = \frac{a_1}{3 \times 4}$$

$$a_5 = \frac{a_2}{4 \times 5} = 0$$

$$a_6 = \frac{a_3}{5 \times 6} = \frac{a_0}{2 \times 3 \times 5 \times 6}$$

$$a_7 = \frac{a_4}{6 \times 7} = \frac{a_1}{3 \times 4 \times 6 \times 7}$$

$$a_8 = \frac{a_5}{7 \times 8} = 0$$

$$a_9 = \frac{a_6}{8 \times 9} = \frac{a_0}{2 \times 3 \times 5 \times 6 \times 8 \times 9}$$

$$a_{10} = \frac{a_7}{9 \times 10} = \frac{a_1}{3 \times 4 \times 6 \times 7 \times 9 \times 10}$$

From here we can proceed in two ways.

1. The first way is to look at the a_n equations that equal either 0, Ca_0, or Da_1. By a proof by induction we can come up with three equations. Either

$$a_{3n} = \frac{a_0}{(3n)(3n-1)(3n-3)(3n-4)...(6)(5)(3)(2)},$$

$$a_{3n+1} = \frac{a_1}{(3n+1)(3n)(3n-2)(3n-3)...(7)(6)(4)(3)},$$

$$a_{3n+2} = 0.$$

Putting these back into our summation for y we get

$$y = a_0 \left[1 + \sum_{n=1}^{\infty} \frac{x^{3n}}{(3n)(3n-1)(3n-3)(3n-4)...(6)(5)(3)(2)} \right]$$

$$+ a_1 \left[x + \sum_{n=1}^{\infty} \frac{x^{3n+1}}{(3n+1)(3n)(3n-2)(3n-3)...(7)(6)(4)(3)} \right].$$

2. The second way is to impose two conditions, $a_0 = 0$ with $a_1 = 1$, and $a_0 = 1$ with $a_1 = 0$. Substituting these with the general power series gives us the same result as in the first way.

Example: (Hermite's differential equation) Solve the differential equation below by power series.

$$\frac{d^2 y}{dx^2} - 2x \frac{dy}{dx} + \lambda y = 0$$

Solution:

Taking y, $\frac{dy}{dx}$, and $\frac{d^2 y}{dx^2}$ as above and substituting into Hermite's equation we get

$$\sum_{n=2}^{\infty} n(n-1)a_n x^{n-2} - 2 \sum_{n=1}^{\infty} na_n x^n + \lambda \sum_{n=0}^{\infty} a_n x^n = 0.$$

Letting $k = n - 2$, $k = n$, and $k = n$ for the first, second, and third summation respectively we get

$$\sum_{k=0}^{\infty} (k+2)(k+1)a_{k+2} x^k - 2 \sum_{k=1}^{\infty} ka_k x^k + \lambda \sum_{k=0}^{\infty} a_k x^k = 0.$$

Notice that the first and last summation start at $k = 0$ and the second starts at $k = 1$. This means we have to pull out the $k = 0$

part from the first and last summation. Doing this we get

$$2a_2 + \lambda a_0 + \sum_{k=1}^{\infty} \left[(k+2)(k+1)a_{k+2} + (\lambda - 2k)a_k \right] x^k = 0.$$

As we stated before, either the terms outside the summation have to equal zero or the term multiplying the x^k have to be zero. From this we generate these two recurrence relations

$$a_2 = -\frac{\lambda}{2} a_0$$

$$a_{k+2} = \frac{2k - \lambda}{(k+2)(k+1)} a_k.$$

Writing these out for $k = 1, 2, \ldots$ we get

$$a_2 = -\frac{\lambda}{2} a_0$$

$$a_3 = \frac{2 - \lambda}{3 \times 2} a_1$$

$$a_4 = \frac{4 - \lambda}{4 \times 3} a_2 = -\frac{(4 - \lambda)\lambda}{4 \times 3 \times 2} a_0$$

$$a_5 = \frac{6 - \lambda}{5 \times 4} a_3 = \frac{(6 - \lambda)(2 - \lambda)}{5 \times 4 \times 3 \times 2} a_1$$

$$\vdots$$

Putting these back into the power series and separating into two solutions y_1 and y_2 we get

$$y_1 = a_0 \left[1 - \frac{\lambda}{2!} x^2 - \frac{(4 - \lambda)\lambda}{4!} x^4 - \frac{(8 - \lambda)(4 - \lambda)\lambda}{6!} x^6 - \ldots \right]$$

$$y_2 = a_1 \left[x + \frac{(2 - \lambda)}{3!} x^3 + \frac{(6 - \lambda)(2 - \lambda)}{5!} x^5 + \ldots \right].$$

6.3 Exercises

1. Change the index of the summation so it starts at $n = 3$

$$\sum_{n=0}^{\infty}(n+1)x^{n+1}$$

2. Show that

$$\sum_{n=0}^{\infty}(1+n^2)x^n \neq 1 + 2x + 3x^2 + \sum_{n=3}^{\infty}(1+n^2)x^n$$

   and explain the mistake.

3. Change the index of the summation of the following expression and simplify into one summation.

$$\sum_{n=1}^{\infty}a_n x^{n-1} + \sum_{n=0}^{\infty} n a_n x^n.$$

4. Solve the differential equation using the power series method.

$$\frac{dy}{dx} - y = 0.$$

5. Solve the differential equation using the power series method.

$$\frac{d^2 y}{dx^2} + y = 0.$$

Chapter 7

Systems of ODEs

So far we have looked at singular equations where we have one independent and one dependent variable. The next step is to look at multiple equations each containing y_1 to y_n which are dependent on x. When theses equations contain the same y_i then they are called a system of ODEs. Below is an example of a system of differential equations where we have y_1 and y_2 and their derivatives.

$$y_1' = y_1 + 3y_2$$
$$y_2' = 2y_1 - y_2$$

Here we have used the shorter Lagrange notation for the derivative. We will continue this notation for the rest of the chapter.

7.1 Eigenvalues and Eigenvectors

Before we can solve systems of ODEs we need to understand more about eigenvalues and eigenvectors. As we shall see in the next section, systems of ODEs can be represented by matrices. This is where eigenvalues and eigenvectors come into play. If we have a vector equation

$$\mathbf{A}x = y$$

where $\mathbf{A}$ is a linear transformation that turns a vector x into a vector y. What we will be interested in is whether we can find a scalar $\lambda \neq 0$ such that

$$\mathbf{A}x = \lambda x. \qquad (7.1)$$

In this case we call the vector x and eigenvector and if any λ exist they are called eigenvalues. These are useful as we can take a matrix and turn it into a scalar.

We can rewrite this eigenvalue equation as

$$(\mathbf{A} - \lambda I)x = 0 \qquad (7.2)$$

where I is the identity matrix. We have to invoke Cramer's rule:

Theorem 8: (Cramer's rule) An $n \times n$ system has a nontrivial solution provided the determinant is non-zero.

This states that if our equation has a solution then $(\mathbf{A} - \lambda I)^{-1}$ exists. Another way of saying this is

$$det(\mathbf{A} - \lambda I) = 0.$$

Any λ that satisfies this are the eigenvalues described above.

Example: Find the eigenvalues of

$$\mathbf{A} = \begin{bmatrix} 3 & -4 \\ 1 & -2 \end{bmatrix}.$$

Solution:

To find the eigenvalues we subtract λI from the matrix $\mathbf{A}$.

$$\mathbf{A} - \lambda I = \begin{bmatrix} 3 - \lambda & -4 \\ 1 & -2 - \lambda \end{bmatrix}$$

Next, taking the determinant of this and letting it equal to zero.

$$det \begin{bmatrix} 3 - \lambda & -4 \\ 1 & -2 - \lambda \end{bmatrix} = 0$$

giving $(3 - \lambda)(-2 - \lambda) - (-4)(1) = 0$. Solving for λ give $\lambda = -1$ and $\lambda = 2$.

If we look at our eigen-equation 7.1 above we see that we know what $\mathbf{A}$ and λ are but we currently do not know what our eigenvector x is. To calculate this we need to substitute our values of λ we have found into the rearranged eigenvalue equation 7.2 and solve.

Example: Find the eigenvectors of

$$\mathbf{A} = \begin{bmatrix} 3 & -4 \\ 1 & -2 \end{bmatrix}.$$

Solution:

We have already found the eigenvalues for this matrix in the previous example. These were $\lambda = -1$ and $\lambda = 2$. Now we need to use these one at a time to find the corresponding eigenvectors.

Let $\lambda = 2$. Then

$$\begin{bmatrix} 3 - 2 & -4 \\ 1 & -2 - 2 \end{bmatrix} \begin{bmatrix} x_1 \\ x_2 \end{bmatrix} = 0$$

$$\begin{bmatrix} 1 & -4 \\ 1 & -4 \end{bmatrix} \begin{bmatrix} x_1 \\ x_2 \end{bmatrix} = 0.$$

As we saw from the previous chapter we can use Gaussian elimination on this to solve for x_1 and x_2.

$$\begin{bmatrix} 1 & -4 & | & 0 \\ 1 & -4 & | & 0 \end{bmatrix}$$

Taking away four lots of row 1 away from row 2 give

$$\begin{bmatrix} 1 & -4 & | & 0 \\ 0 & 0 & | & 0 \end{bmatrix}$$

Writing this back out as equation gives

$$x_1 - 4x_2 = 0$$

Setting $x_2 = 1$ we see that $x_1 = 4$. Putting these into vector form our eigenvector is

$$\begin{bmatrix} x_1 \\ x_2 \end{bmatrix} = \begin{bmatrix} 4 \\ 1 \end{bmatrix}.$$

This is the eigenvector associated to the specific eigenvalue $\lambda = 2$. We have to do this for all the eigenvalues we have obtained so next we let $\lambda = -1$ and repeat this process.

$$\begin{bmatrix} 3 - -1 & -4 \\ 1 & -2 - -1 \end{bmatrix} \begin{bmatrix} x_1 \\ x_2 \end{bmatrix} = 0$$

$$\begin{bmatrix} 4 & -4 \\ 1 & -1 \end{bmatrix} \begin{bmatrix} x_1 \\ x_2 \end{bmatrix} = 0.$$

Using Gaussian elimination and taking four lots of row 2 away from row 1 we get

$$\begin{bmatrix} 0 & 0 & | & 0 \\ 1 & -1 & | & 0 \end{bmatrix}$$

Writing this back out as equation gives

$$x_1 - x_2 = 0$$

so we can straight away see that $x_1 = x_2$. Setting $x_1 = 1$ we get the eigenvector as

$$\begin{bmatrix} x_1 \\ x_2 \end{bmatrix} = \begin{bmatrix} 1 \\ 1 \end{bmatrix}.$$

7.2 Systems of ODEs

Like with systems of linear equations we can write a system of equations in matrix form. Let's look at the example given at the start of the chapter. We can represent it in the following way:

$$\begin{bmatrix} y_1' \\ y_2' \end{bmatrix} = \begin{bmatrix} 1 & 3 \\ 2 & -1 \end{bmatrix} \begin{bmatrix} y_1 \\ y_2 \end{bmatrix}.$$

In general here we have a vector $\mathbf{Y}'$ equalling a matrix $\mathbf{A}$ multiplied by a vector $\mathbf{Y}$

$$\mathbf{Y}' = \mathbf{AY}. \tag{7.3}$$

As it is we struggle to solve this however we have already seen a very important statement that can help us with this. As we saw above, we can look for the eigenvalues and eigenvectors and turn this equation into one where instead of having a matrix $\mathbf{A}$ we have a scalar λ. This is important as scalars are a lot easier to deal with than scalars. Like with the earlier theory we assume that $y = \theta e^{mx}$, where θ is a vector θ_1 to θ_n, is a solution. Substituting this into 7.3 above we get

$$m\theta e^{mt} = \mathbf{A}\theta e^{mt}$$

or

$$(\mathbf{A} - mI)\theta e^{mx} = 0$$

Remember that $e^{mx} \neq 0$ so $(\mathbf{A} - mI)\theta = 0$. This is like our eigenvalue equation above so we can work out the eigenvalues and corresponding eigenvectors for these but instead of using λ we use m.

Example: Solve the system of differential equations below.

$$y_1' = 3y_1 - 4y_2$$
$$y_2' = y_1 - 2y_2$$

Solution:

We start by putting this into matrix form

$$\begin{bmatrix} y_1' \\ y_2' \end{bmatrix} = \begin{bmatrix} 3 & -4 \\ 1 & -2 \end{bmatrix} \begin{bmatrix} y_1 \\ y_2 \end{bmatrix}.$$

Next we need to let $\mathbf{Y} = \theta e^{mx}$ and substitute this and its derivative into our matrix equation above.

$$\begin{bmatrix} m\theta_1 e^{mx} \\ m\theta_2 e^{mx} \end{bmatrix} = \begin{bmatrix} 3 & -4 \\ 1 & -2 \end{bmatrix} \begin{bmatrix} \theta_1 e^{mx} \\ \theta_2 e^{mx} \end{bmatrix}$$

Making one side of this equation equal to zero we obtain

$$\begin{bmatrix} 3-m & -4 \\ 1 & -2-m \end{bmatrix} \begin{bmatrix} \theta_1 e^{mt} \\ \theta_2 e^{mt} \end{bmatrix} = \begin{bmatrix} 0 \\ 0 \end{bmatrix}.$$

As we have worked out the eigenvalues and eigenvectors for this already in an example previously we will not go through the written calculations again. From above we have these to be

$$\begin{bmatrix} y_1 \\ y_2 \end{bmatrix} = \begin{bmatrix} 4 \\ 1 \end{bmatrix} \text{ for } \lambda = 2,$$

$$\begin{bmatrix} y_1 \\ y_2 \end{bmatrix} = \begin{bmatrix} 1 \\ 1 \end{bmatrix} \text{ for } \lambda = -1.$$

Since we have values for λ and the eigenvectors associated with these, we can substitute them into our suggested solution $\mathbf{Y} = \theta e^{mx}$ giving two solutions.

$$\mathbf{Y}_1 = \begin{bmatrix} 4 \\ 1 \end{bmatrix} e^{2x}$$

$$\mathbf{Y}_2 = \begin{bmatrix} 1 \\ 1 \end{bmatrix} e^{-x}$$

Note that there is a difference between $\mathbf{Y}_1$ and y_1. y_1 is an element in the vector $\mathbf{Y}$ where as $\mathbf{Y}_1$ is the first solution to the system of ODEs. To obtain $\mathbf{Y}$ we need to sum up all our solutions as we did in previous chapters.

$$\mathbf{Y} = \begin{bmatrix} y_1 \\ y_2 \end{bmatrix} = c_1 \begin{bmatrix} 4 \\ 1 \end{bmatrix} e^{2x} + c_2 \begin{bmatrix} 1 \\ 1 \end{bmatrix} e^{-x}$$

where we have used c_1 and c_2 as constants. If we take each row of the vectors we come up with two equations for the solutions, one for y_1 and one for y_2:

$$y_1 = 4c_1 e^{2x} + c_2 e^{-x},$$
$$y_2 = c_1 e^{2x} + c_2 e^{-x}.$$

7.3 Exercises

1. Find the eigenvalues and eigenvectors of

$$\begin{bmatrix} -2 & 1 \\ 12 & -3 \end{bmatrix}.$$

2. By inspection only what are the eigenvalues of

$$\begin{bmatrix} 5 & 0 & 0 \\ 0 & -3 & 0 \\ 0 & 0 & 1 \end{bmatrix}.$$

3. Find the eigenvalues of

$$
\begin{bmatrix}
-2 & -4 & 2 \\
-2 & 1 & 2 \\
4 & 2 & 5
\end{bmatrix}.
$$

4. Write the following system of equations in matrix form.

$$
\begin{aligned}
y_1' &= 3y_1 - 2y_2 \\
y_2' &= -2y_1 + 5y_2
\end{aligned}
$$

5. Find the general solution for the system of differential equations below.

$$
\begin{aligned}
y_1' &= y_1 + 2y_2 \\
y_2' &= 2y_1 + y_2.
\end{aligned}
$$

6. Find the general solution for the system of differential equations below.

$$
\begin{aligned}
y_1' &= y_1 - 3y_2 \\
y_2' &= 4y_1 - 6y_2.
\end{aligned}
$$

Solutions

Chapter 1

1. Find the gradient of the line that passes through the points (3,4) and (6,5).

$$m = \frac{1}{3}$$

$$y - 5 = \frac{1}{3}(x - 6)$$

2. Find the gradient of the line that passes through the points (6,1) and (2,9).

$$m = -2$$

$$y = 13 - 2x$$

3. Is there a single straight line that passes through (2,1), (-1,0), and (5,4)?

No. The equation for the line that passes through (2,1) and (-1,0) is $y = x/3 + 1/3$. (5,4) Does not appear on this line.

4. Find the equation of the tangent to $y = 2/x$ at $x_0 = 2$.

$$m = \lim_{x \to 2} \frac{\frac{2}{x+h} - 1}{h}$$

$$m = \lim_{x \to 2} \frac{-h}{h(2 + h)}$$

$$m = \lim_{x \to 2} \frac{-1}{(2 + h)} = -1/2$$

Substituting m and the point point (2,1) we obtain $y = -1/2x + 2$

5. Prove that the derivative of e^x is e^x

$$(e^x)' = \lim_{h \to 0} \frac{e^{x+h} - e^x}{h}$$

$$(e^x)' = \lim_{h \to 0} \frac{e^x(e^h - 1)}{h}$$

$$(e^x)' = e^x \lim_{h \to 0} \frac{e^h - 1}{h} = e^x$$

6. Use your answer to question 3 and a substitution of $u = nx$ to prove that the derivative of e^{nx} is ne^{nx}

   Do the same as above but substitute $u = nx$ first.

7. $\frac{dy}{dx} = 24x^2(2x^3 + 5)^3$

8. $\frac{dy}{dx} = 6e^{2x}$

9. $\frac{dy}{dx} = -\frac{1}{2y^2 + sin(y)}$

10. $\frac{dy}{dx} = \frac{sin(x) + 2x}{4 + cos(y)}$

11. $\frac{1}{3}x^2e^{3x} - \frac{2}{9}xe^{3x} + \frac{2}{27}e^{3x} + c$

12. $\frac{1}{3}x^3 ln|x| - \frac{1}{9}x^3 + c$

Chapter 2

1. (i) First order linear equation, initial value. (ii) First order non-linear equation, initial value. (iii) Third order linear equation, initial value. (iv) Second order non-linear equation, boundary value.

2. (i) False, (ii) False , (iii) True.

3. $y = A - \frac{x^3}{3} + 2x$

4. $y = Ae^{-\frac{3x}{c}}$

5. General solution $y = e^{5x}$, full solution $y = 2e^{5x}$

6. General solution $y = Ae^{x^3}$, full solution $y = 4e^{x^2}$

7. General solution $y = Ce^{sin(x)} - 1$, full solution $y = 2e^{sin(x)} - 1$

8. $y = \frac{A}{x} + sin(x) + \frac{cos(x)}{x}$

9. $y = \frac{a}{Ax - ln(x)}$

10. $y = 2e^{2x} + \frac{A}{e^x}$

11. $y = x^2(A - cos(x))$

Chapter 3

1. $y = Ae^x + Be^{-5x}$

2. $y = A\sin(\sqrt{5}x) + B\cos(\sqrt{5}x)$

3. General solution $y = Ae^{-2x} + Be^{-x}$, full solution $y = -6e^{-2x} + 9e^{-x}$.

4. General solution $y = A + Be^{-5}$, full solution $y = \frac{11}{5} - \frac{e^{-x}}{5}$.

5. General solution $y = Ae^{\frac{-x}{2}}\sin(\frac{3x}{2}) + Be^{\frac{3x}{2}}\cos(\frac{3x}{2}) + \frac{3}{13}\sin(x) - \frac{2}{13}\cos(x)$, full solution $y = -\frac{4}{39}e^{\frac{-x}{2}}\sin(\frac{3x}{2}) + \frac{2}{13}e^{\frac{3x}{2}}\cos(\frac{3x}{2}) + \frac{3}{13}\sin(x) - \frac{2}{13}\cos(x)$

6. General solution $y = A\sin(2x) + B\cos(2x) + \frac{x^3}{4} - \frac{3x}{8}$, full solution $y = \frac{3}{16}\sin(2x) + 2\cos(2x) + \frac{x^3}{4} - \frac{3x}{8}$

7. General solution $y = \frac{A}{x^2} + \frac{B}{x^3}$, full solution $y = \frac{18}{x^2} - \frac{14}{x^3}$

8. $y = Ae^{-2x}\sin(3x) + Be^{-2x}\cos(3x)$

9. General solution $y = Axe^{-2x} + Be^{-2x}$

10. $y = Acos(3x) + Bsin(3x) + \frac{1}{9}(1 + 2x)$

11. General solution $y = A + Bln(x)$, full solution $y = 3 + 4ln(x)$

Chapter 4

1. $-\frac{5}{x-2} + \frac{2}{3x} + \frac{13}{3(x-3)}$

2. Use integration by parts twice similar to example in chapter one then solve for $\mathcal{L}[sin(at)]$.

3. $\frac{1}{s-5} - \frac{4s}{s^2+4} + \frac{24}{(s-3)^3}$

4. Laplace transform $s^2Y(s) + Y(s) = \frac{2}{s^2+1}$, full solution $y = \frac{1}{4}e^{-t} + \frac{1}{4}e^x - \frac{1}{2}cos(t)$

5. Substituting e^t into the Laplace transform we see that $F(s) = [\frac{1}{s-1}e^{(1-s)t}]_0^\infty$ giving the desired result.

6. Laplace transform $sY(s) - 1 + Y(s) = \frac{1}{s^2}$, full solution $y = x + 2e^{-x} - 1$

7. Laplace transform $(s^2 - 3s + 2)Y(s) - 3 + 3 = \frac{1}{s-3}$, full solution $y = \frac{5}{2}e^t - 2e^{2t} + \frac{1}{2}e^{3t}$

8. Laplace transform $sY(s) + 5 + Y(s) = \frac{1}{s-3}$, full solution $y = -6e^{2t} + e^{3t}$

9. Laplace transform $sY(s) - 3 = \frac{1}{(s-1)^2} + \frac{2}{s-1} + Y(s)$, full solution $y = 3e^x + 2xe^x + \frac{x^2}{2}e^x$

Chapter 5

1. In matrix form $\begin{bmatrix} 1 & 0 & 3 & -3 \\ 0 & 1 & 1 & -2 \\ 2 & 2 & 2 & 2 \end{bmatrix}$ and solving gives $A = 3$, $B = 2$ and, $C = -4$.

2. $y = Ae^{-x} + Be^x + Cxe^x$

3. $y = Ae^x + Be^x + Ce^{2x}$

4. General solution $Ae^{-x} + Be^{-x} + Cx^2e^{-x}$, full solution $y = \frac{5}{2}x^2e^{-x}$

157

5. General solution $y = Ae^{-x} + Be^{-x} + Ce^{x} + Dxe^{x}$, full solution $y = xe^{x} + e^{x}$

6. $y = Ae^{-5x} + Be^{-x} + Ce^{x} + Dxe^{x} + Fxe^{-x}$

7. $y = \frac{A}{6}sin(x) - Bcos(x) + C - sin(2x) + 1$

8. $y = Acos(x) + Bsin(x) + C + \frac{e^{x}}{2}$

Chapter 6

1. $2x^2 + x + \sum_{n=2}^{\infty}(n+1)x^{n+1}$.

2. Writing out the first summation give until the index is $n = 3$ gives $1 + 2x + 5x^2 + \sum_{n=3}^{\infty}(1+n^2)x^n$. The n term in the bracket has not been squared on the right hand side.

3. $\sum_{k=0}^{\infty}\left[a_{k+1} + ka_k\right]x^k$.

4. We see that $a_n = a_0\frac{1}{n!}$ giving $y = \sum_{n=0}^{\infty}a_0\frac{x^n}{n!} = a_0e^{x}$

5. $y = a_0 \sum_{n=0}^{\infty} \frac{(-1)^n x^{2n}}{(2n)!} + a_1 \sum_{n=0}^{\infty} \frac{(-1)^n x^{2n+1}}{(2n+1)!}$

Chapter 7

1. Eigenvalues $\lambda = -6$ and $\lambda = 1$ with the corresponding eigenvectors

$$\begin{bmatrix} -1 \\ 4 \end{bmatrix} \text{ and } \begin{bmatrix} 1 \\ 3 \end{bmatrix}.$$

2. As there is zeros in all the elements of the matrix apart from the diagonal, the eigenvalues are these values on the diagonal.

3. The eigenvalues are $\lambda = 3$, $\lambda = -5$ and $\lambda = 6$.

4. The matrix form of the system is

$$\mathbf{Y} = \begin{bmatrix} y_1' \\ y_2' \end{bmatrix} = \begin{bmatrix} 3 & -2 \\ -2 & 5 \end{bmatrix} \begin{bmatrix} y_1 \\ y_2 \end{bmatrix}.$$

5. The eigenvalues are $\lambda = 3$ and $\lambda = -1$ with respective eigenvectors $\begin{bmatrix} 1 \\ 1 \end{bmatrix}$ and $\begin{bmatrix} -1 \\ 3 \end{bmatrix}$ giving the general solution to be $\mathbf{Y} = c_1 \begin{bmatrix} 1 \\ 1 \end{bmatrix} e^{3t} + c_2 \begin{bmatrix} -1 \\ 3 \end{bmatrix} e^{-t}.$

6. The eigenvalues are $\lambda = -3$ and $\lambda = -2$ with respective eigenvectors $\begin{bmatrix} \frac{3}{4} \\ 1 \end{bmatrix}$ and $\begin{bmatrix} 1 \\ 1 \end{bmatrix}$ giving the general solution to be

$$\mathbf{Y} = c_1 \begin{bmatrix} \frac{3}{4} \\ 1 \end{bmatrix} e^{-3t} + c_2 \begin{bmatrix} 1 \\ 1 \end{bmatrix} e^{-2t}.$$

Index

Printed in Great Britain
by Amazon

85360468R00098